AF559655

Procedures for Analyzing Chemical Quality of Milk and Milk Products

NIPA® GENX ELECTRONIC RESOURCES & SOLUTIONS P. LTD.
New Delhi-110 034

About the Authors

Dr. Soma Maji is currently working as an Assistant Professor at Centurion University of Technology and Management in Paralakhemundi, Odisha, in the Department of Dairy Technology, School of Agricultural and Bio Engineering. She graduated with a B.Tech in Dairy Technology, an M.Tech in Dairy Chemistry from West Bengal University of Animal and Fishery Sciences in Nadia, West Bengal, and a Ph.D. in Dairy Chemistry from the National Dairy Research Institute in Karnal, Haryana. She has published more than 15 research, technical, review articles and popular articles. She has won many awards and participated in numerous national and international seminars, conferences, training sessions, and workshops. She also received an institute fellowship for her Ph.D. and merit scholarships for her B.Tech and M.Tech studies. Dr. Soma is the author of a handbook on supercritical carbon dioxide extraction and has contributed chapters to several books published by domestic and foreign publishers.

Dr. Pinaki Ranjan Ray is currently employed with West Bengal University of Animal and Fishery Sciences in Kolkata as a professor and head of the Department of Dairy Chemistry. He has been conducting research and teaching in the field of dairy chemistry for over 23 years. In addition to mentoring numerous postgraduate students, he has over 60 research papers and technical articles published in reputable national and international journals. He received several honours and went to many national and international conferences in India and other countries. Dr. Ray is the author of one dairy chemistry textbook and has contributed chapters to several books published by domestic and foreign publishers. He is a lifelong member of SASNET and the Dairy Technology Society of India

Advances in Processing Preservation and Value Addition Technologies Volume 03

Procedures for Analyzing Chemical Quality of Milk and Milk Products

Soma Maji
Assistant Professor
Department of Dairy Technology
School of Agricultural and Bio Engineering
Centurion University of Technology and Management
Paralakhemundi, Odisha

Pinaki Ranjan Ray
Professor and Head
Department of Dairy Chemistry
West Bengal University of Animal and Fishery Sciences
Kolkata, West Bengal

NIPA® GENX ELECTRONIC RESOURCES & SOLUTIONS P. LTD.
New Delhi-110 034

**NIPA® GENX ELECTRONIC
RESOURCES & SOLUTIONS P. LTD.**

101,103, Vikas Surya Plaza, CU Block
L.S.C.Market, Pitam Pura, New Delhi-110 034
Ph : +91 11 27341616, 27341717, 27341718
E-mail:newindiapublishingagency@gmail.com
www: www.nipabooks.com

For customer assistance, please contact
Phone: + 91-11-27 34 17 17
Fax: + 91-11-27 34 16 16

ISBN: 978-93-95763-83-7

Composed and Designed by NIPA.

Preface

Milk and its diverse array of derivative products have a rich and enduring history in human nutrition. The significance of analysing milk and milk products cannot be emphasised, both historically and in the context of dairy farming today. Milk is a comples food that includes superior macromolecules such as fat, protein, and carbohydrates. They are crucial in determining and preserving the stability and quality of milk and milk products. This manual's main goal is to offer helpful guidelines for evaluating milk and milk products' quality at the laboratory scale, along with a range of traditional methods for analysing them. This comprehensive guide on milk and milk product analysis is intended to serve as a useful tool for experts, scholars, and students working in the area of dairy science and technology. The goal of this book is to offer useful guidance on the analysis of different types of milk and milk-based products that adheres to legal requirements for finished goods while covering a broad range of parameters. The manual includes information on analysing raw milk, including its chemical makeup and physical characteristics, as well as analysis techniques for spotting adulteration in milk and fat-rich dairy products like cheese, ice cream, and dried milk. Additionally, this book covers how to prepare various standard solutions and calibrate glassware used in analytical processes.

The authors acknowledge the support rendered by Dr. Anamika Das and Mr. Mitul Kumar Bumbadiya in the preparation of this book. The authors also acknowledge the support from the Centurion University of Technology and Management, Odisha and West Bengal University of Animal and Fishery Sciences, Mohonpur, Nadia. W.B.

Authors

Contents

1

Analysis of Raw Milk

1 Analysis of Chemical Composition of Milk

1.1 Determination of Acidity of Milk

Introduction

The acidity of milk can be determined by titrating the milk against standard sodium hydroxide solution using an indicator like phenolphthalein and is expressed in the terms of lactic acid. Though natural acidity in milk is due to its constituents such as casein, albumin, citrates, phosphates and carbon dioxide but it is customary to express in terms of lactic acid.

As one gram equivalent weight of an alkali neutralizes one gram equivalent weight of an acid, the volume of standard alkali required to neutralize lactic acid of fixed volume of milk gives percent acidity of milk according to the reaction.

$CH_3\text{-}CHOH\text{-}COOH + NaOH \rightarrow CH_3CHOH\text{-}COONa + H_2O$

Lactic acid Sodium Hydroxide Sodium lactate water

Phenolphthalein used as an indicator changes from its colourless form at pH 6.5 of milk to pink colour form at its end point of pH 8.3 and is a fair indicator of acidity of milk. Pink colour end point of phenolphthalein with pink colour of rosaniline acetate in blank determination.

Materials Required

Reagents: NaOH-0.1(N) Phenolphthalein indicator: 0.5%(w/v) in Ethyl alcohol (95% by volume) milk sample

Glasswares: Burette, Burette stand, pipette, porcelain dish

Procedure

1. Thoroughly mix the milk avoiding incorporation of air bubbles.
2. Measure accurately 10 ml of milk in two porcelain dishes.
3. Add an equal volume of freshly boiled and cooled water.

4. Add 1 ml of phenolphthalein indicator solution to one of the dish and to the other add 1 ml of bench solution of rosaniline acetate.
5. Titrate the contents of the dish to which phenolphthalein has been added, against 0.1 N standard sodium hydroxide solution added drop by drop from the burette
6. Compare the colour matches the pink tint of the control sample.
7. The time taken to complete titration should not exceed 20 seconds.

Observation

Volume of standard sodium hydroxide solution required for titration (ml) (V_1) = _____

Volume of milk taken for titration (ml) (V_2) = _____

Normality of the standard sodium hydroxide solution = _____ N

Calculation

$$\text{Titratable acidity (\% lactic acid)} = \frac{9V_1N}{V_2}$$

Result

The given sample of milk has _____% (lactic acid) titratable acidity.

1.2 Determination of pH of milk

Introduction

Cow's milk is acidic to phenolphthalein, alkaline to methyl orange but amphoteric to litmus due to the presence of phosphates. The pH of milk is usually between 6.4 and 6.6. The electrometric measurement using pH meters is now relatively simple and accurate. It measures the potential difference between the glass electrode and a standard calomel electrode, which can form parts of a combination electrode, and are calibrated by the use of prepared buffer solutions of accurately known pH.

Apparatus: pH meter with electrode, Beaker

Reagents: Standard buffer solutions (pH 4, 7, 9)

Procedure

1. About 20 ml of thoroughly mixed milk sample was taken into a 50 ml beaker.

2. The pH meter was first standardized using standard buffer solution (pH 4, pH 7 and pH 11)
3. Then the pH of milk sample was measured directly on the pH meter at a fix temperature.

Observation

pH value as displayed on pH meter =_____

Result

The given sample of milk has a pH of _____.

1.3 Determination of Fat Content of Milk

1.3.1 Gerber Method

Introduction

In Gerber method, sulphuric acid is added to milk in order to dissolve the protein and other solids and create heat which liquefy and de-emulsify the fat. Iso-amyl alcohol is added to lower down the surface tension and aid the separation of fat. Due to centrifugation, fat being lighter, will separate on the top of solution in the milk butyrometer. It is volumetric method.

Apparatus

Milk butyrometer, Stopper and key, Butyrometer stand, Centrifuge machine, Water bath

Reagents

- Gerber acid- 90% (v/v) (Sp. gravity - 1.007-1.812 at 20ºC)
- Iso- Amyl alcohol- (Sp. gravity 0.803 to 0 .805 at 27ºC)

Procedure

1. Take 10 ml of Gerber acid into milk butyrometer with the help of tilt measure.
2. Pipette out 10.75 ml of the well mixed sample of milk and transfer it to the butyrometer carefully without allowing it to mix with the acid.
3. Add 1 ml amyl alcohol with the help of tilt measure.
4. Close the butyrometer with rubber stopper and mix the content by shaking until all the curd has been digested.

5. Transfer the butyrometer in water bath at 65±1°C for 5 min.
6. Place the butyrometer in the centrifuge, balance the machine by placing another butyrometer filled with milk or water to the opposite side and centrifuge for 5 min at 1100 rpm.
7. Transfer the butyrometer in a water bath, keeping the stoppers downwards, at 65±1°C for 3 min.
8. Adjust the fat column within the scale on butyrometer and take reading.

Observation

Fat percent (%) as displayed on the stem of the milk butyrometer =_____

Result

The fat content of the given sample of milk is _____ %

1.3.2 Rose-Gottlieb Method

Introduction

The milk sample is treated with ammonia and ethyl alcohol. Ammonia is used to dissolve the protein and ethyl alcohol to precipitate the proteins by reducing the surface tension of the solution. Fat is extracted with diethyl ether and petroleum ether. Mixed ethers are evaporated and the residue is weighed. This method is considered suitable for reference purposes. It is gravimetric method.

Apparatus

Rose Gottlieb tubes, siphon assembly, conical flask, water bath, hot air oven, desiccator, weighing balance

Reagents

- Ammonia (Sp. gravity 0.8974 at 16°C)
- Petroleum Ether (Sp. gravity – 0.065, boiling point 40 – 60°C)
- Di ethyl ether (Sp. gravity – 0.720, Peroxide free)
- Ethyl alcohol - 95%

Procedure

1. Weigh accurately by difference using a weighing bottle, 10 ml of the well mixed sample of milk into the extraction tube.
2. Add 1 ml concentrated ammonia and mix well by gentle swirling action.

3. Add 10 ml of alcohol and again mix well.
4. Add 25 ml of dicthyl ether, close the tube with cork or stopper which should be wetted with water and shake vigorously for 1 min.
5. Remove the cork, wash the cork and neck of the tube with 25 ml of light petroleum ether in such a way that the washing run into the tube.
6. Insert the cork again and shake vigorously for 30 seconds.
7. Allow the tube to stand until the ether layer is clearly separated from the aqueous layer, usually not less than 30 min.
8. Remove the cork and insert the siphon fitting so adjusted for length that the inlet is 2-3 mm above the interface between the ethereal and aqueous layer.
9. Transfer the ethereal layer to a suitable clean, dry and previously weighed flask.
10. Remove the siphon fitting and repeat the extraction of the milk residue again two times using 15 ml of petroleum ether and 15 ml of di ethyl ether as before.
11. Evaporate the solvents using a hot plate (100°C) from the flask and dry the residual fat in the oven at 100 ± 1°C for 1 h.
12. Cool the flask in an efficient desiccator and weigh.
13. Repeat the process of heating, cooling and weighing at an interval of 30 min until successive weighing do not show a loss in weight by more than 0.1 mg.

Observation

W1= Weight of empty flask

W2= Weight of flask after drying

W= Weight of sample

Calculation

$$\text{Fat in milk (\% by wt.)} = \frac{(w_2 - w_1)}{w} \times 100$$

Result

The fat content of the given sample of milk is _____ %

1.4 Determination of Total Solids and Solids not Fat Content in Milk

1.4.1 Gravimetric Method

Introduction

A known quantity of milk is dried on a boiling water bath. Subsequently sample is dried in hot air oven at 100 ± 1°C and from the weight of the residue, the total solids content in milk is determined.

Apparatus

Hot air oven, Desiccator, Aluminum moisture dish, weighing balance, glass pipette

Procedure

1. Weigh accurately a clean, dry empty dish with glass rod.
2. Pipette 5 ml of the well mixed sample of milk into the dish and weigh it accurately.
3. Place the dish on a boiling water bath for at least 30 min.
4. Remove the dish, wipe the bottom and transfer to a well-ventilated oven at 100±1°C.
5. After 3 h transfer the dish immediately into a desiccator, cool it to room temperature and weigh.
6. Repeat the process of heating, cooling and weighing until the loss of weight between two successive weighing does not exceed 0.5 mg. Note the lowest weight.

Observation

W2 = Weight of dish after drying

W1 = Weight of empty dish

W = Weight of sample

Calculation

$$\text{Total solids (\% by wt.)} = \frac{(w_2 - w_1)}{w} \times 100$$

SNF (%) = Total solids (%) - Fat (%)

Result

The given sample of milk has total solids content of _____ %

The given sample of milk has SNF content of _____%

1.4.2 Volumetric method (Using Lactometer)

Introduction

Lactometers are used to quickly determine the specific gravity of milk. The method is based on the law of floating, which states that a solid is subjected to an upward thrust when immersed in liquid, equal to the weight of the liquid displaced from the solid and acts vertically upwards. The total solids and SNF can be determined if lactometer reading, temperature of milk and fat content is known.

Apparatus

Lactometer, Lactometer Jar, Thermometer

Procedure

1. Warm the milk sample at 40°C for 5 min.
2. Cool the sample at the temperature at which the lactometer is calibrated to work (e.g. 29°C for Zeal lactometer).
3. Pour enough milk into the cylinder taking care to avoid incorporation of air bubbles.
4. Some milk should overflow when lactometer is immersed.
5. Dip the lactometer gently. The lactometer should float freely and should not touch the side of the cylinder.
6. Allow the lactometer to come to rest in the milk.
7. Take the reading within about 30 s and also note the temperature.
8. Convert the observed lactometer reading to the corrected lactometer reading (CLR) for temperature, if necessary.
9. For this, add 0.1 in the observed reading for each°C above the require temperature and subtract 0.1 from the observed reading for each °C below the required temperature.
10. Determine the fat percentage of milk sample by Gerber's method.
11. Place the figures of fat and CLR in the Richmond's formula for calculating total solids and solids not fat.

Observation

CLR = corrected lactometer reading at 29°C

F= fat % = by Gerber method

Calculation

$$\text{Total Solids (\%)} = \frac{CLR}{4} + 1.2F + 0.14$$

$$\text{SNF (\%)} = \frac{CLR}{4} + 0.2F + 0.14$$

Result

The given sample of milk has total solids content of_____ %

The given sample of milk has SNF content of_____%

1.5 Determination of Protein Content in Milk

1.5.1 Kjeldahl Method

Introduction

Kjeldahl method is the most widely used method for determination of protein. The method involves two major steps. In the first step, the protein is digested using concentrated sulphuric acid in presence of a catalyst and form ammonium sulphate. Neutral salts such as potassium sulphate are used in the digestion step to raise the boiling point of the reaction mixture and metallic catalyst such as copper sulfate is used to hasten the digestion and clearing the reaction mixture. In the second step, the digested material is neutralized with alkali to liberate ammonia and the liberated ammonia is distilled off, collected in boric acid as ammonium borate complex and titrated with standard acid in the presence of a methyl red- methylene blue indicator until the green distillate changes through colour less to pink. The quantity of acid required for titration is equivalent to the concentration of ammonia in the distillate and to the nitrogen content of the original protein containing sample. A reagent blank should be run to subtract reagent nitrogen from sample nitrogen. The protein content comes from this method is represented as crude protein as some non- protein nitrogen is also present in it. The total amount of protein content in milk is calculated using the conversion factor of 6.38.

Reaction

$$\text{Protein} + \xrightarrow{K_2SO_4, CuSO_4\, H_2SO_4} (NH_4)_2SO_4$$

$$(NH_4)_2SO_4 + 2NaOH + \longrightarrow 2NH_3 + Na_2SO_4 + 2H_2O$$

$$NH_3 + H_3BO_3 \longrightarrow NH_4 + H_2BO_4^{-}$$

$$NH_4 + \underset{\text{(Green color)}}{H_2BO_4^{-}} + HCl \longrightarrow \underset{\text{(Pink color, pH<4.8)}}{NH_4Cl + H_3BO_3}$$

Apparatus

Water bath, Analytical balance, Digestion block assembly (Exhaust manifold; Aspirator), Distillation unit, Digestion tube, 250 ml capacity, Automatic pipettes (dispensers), Graduated measuring cylinder 50 ml capacity, Conical flask 250 ml capacity, Burette 25 ml capacity.

Reagents

- Kjeldahl catalyst mixture: It consists of 3.5 g potassium sulphate and 0.105 g copper sulfate.
- Sulfuric acid: with a mass fraction of at least 98% nitrogen free.
- Sodium hydroxide solution: Nitrogen free, containing approximately 40 g sodium hydroxide per 100 ml.
- Boric acid solution: Dissolve 40 g of boric acid in 1 litre of hot water in a 1000 ml one-mark volumetric flask. Allow the contents to cool to 20°C and adjust the mark with water.
- Indicator solution: Dissolve 0.25 g of methylene blue and 0.375 g of methyl red in 300 ml of 95 per cent ethanol.
- Hydrochloric acid standard volumetric solution 0.1N
- Sucrose, with nitrogen content not more than 0.002%.

Procedure

Digestion of sample

1. In a clean and dry Kjeldahl digestion tube, 5 g of digestion mixture and 2 g of milk sample was added.
2. 10 ml of concentrated sulphuric acid along the sides of the digestion tube was added carefully and mixed the contents.
3. The tubes were placed in digestion stand and kept in the digestion unit (maintained at 420°C) and left it for 2 hours,

4. Water supply should be properly connected to the digestion system. After 2 hours the colour of samples turned into bluish green.
5. When bluish green colour appears, the tubes were removed from the block and placed them in cooling stand until the fumes disappeared and the colour becomes crystal clear.

Distillation

1. The digestion tubes were then transferred to distillation unit and subjected to automatic distillation by placing a conical flask containing 50 ml of boric acid solution under the outlet of condenser in such a way that the delivery tube was below the surface of the excess boric acid solution.
2. Distillation unit was adjusted to dispense 40 ml of sodium hydroxide solution and distilled off the ammonia liberated by the addition of sodium hydroxide solution and collected the distillate in the excess boric acid solution.
3. The content of the conical flask was titrated with the standard hydrochloric acid till the appearance of violet colour when presence for at least 15 second in the contents.
4. Similarly blank test was conducted by addition of 0.2 g sucrose in place of milk and titrate the distilled solution against 0.1 N HCl.

Observation

A = ml of standard hydrochloric acid solution used for milk sample

B = ml of standard hydrochloric acid solution used for blank

N = Normality of the standard hydrochloric acid solution

M = mass in g of milk sample

Calculation

$$\text{Spelling Nitrogen (\%)} = \text{Nitrozen (\%)} = \frac{1.4007(A - B)\times N}{\text{M}}$$

Total Protein = Nitrogen (%) x 6.38

Result

The nitrogen content of the given sample of milk is _____ %

The protein content of the given sample of milk is _____ %

1.5.2 Pyne's Method

Introduction

This method involves dual titration. Initially saturated potassium oxalate is used to chelate the calcium present in milk. The first titration is done to neutralize the acidity of milk. After that neutralized formaldehyde added to the mix which reacts with amino (NH_2) group of protein and releases hydrogen ion. The amount of alkali used in the second titration is the measure of the amino groups present in milk. Milk protein is determined by multiplying 1.7 with second titre value.

Apparatus

Burette, Conical flask, funnels

Reagents

- Neutralized formaldehyde solution
- Saturated potassium oxalate
- Phenolphthalein indicator
- Sodium hydroxide solution (0.1N)

Procedure

1. Pipette 10 ml of well mixed sample of milk in 100 ml conical flask,
2. Add 1 ml of phenolphthalein indicator and 0.4 ml of saturated potassium oxalate solution.
3. Keep the flask aside for 3-4 min without disturbing.
4. Titrate the milk against 0.1N sodium hydroxide to its end point.
5. Add 2 ml of neutral formaldehyde mix well.
6. The pink colour disappears.
7. Continue the titration to the same end point and note the value of alkali required in the second titration.

Observation

BR = Burette reading

Calculation

Protein (% by weight) = BR (second titration) X 1.7

Result

The protein content of the given sample of milk is _____ %

1.6 Determination of Lactose Content in Milk by Lane-Eynon Method

Introduction

Lactose is the major carbohydrate in milk (approximately 48-50 g/l in cow milk) of all the mammals. Lane-Eynon method for the estimation of lactose is based on the reducing property of lactose present in milk. Reducing sugars are able to function as reducing agents because of free aldehyde group present in the molecule. They have the to reduce metal ions notably copper, iron or silver in alkaline solution. This very property of sugars has been used in this method using Fehling solution. Fehling solution is a mixture of Fehling A ($CuSO_4$) and Fehling solution B (alkaline sodium-potassium tartrate). When CuSO4 is made alkaline (during mixing of Fehling A and Fehling B), the $Cu(OH)_2$ gets precipitated but in the presence of sodium-potassium tartrate, it forms a soluble complex with copper compound and prevents the precipitation of $Cu(OH)_2$. The complex is soluble and reacts as if it is alkaline $Cu(OH)_2$ solution. The Fehling solution when heated gives rise to cupric oxide (CuO) which in turn reacts with reducing sugar and gets reduced to cuprous oxide (Cu_2O), brick-red precipitates, resulting into the oxidation of sugars to corresponding acids.

Milk sample is treated with acetic acid to precipitate protein and fat. The filtrate so obtained is added drop wise with Alkaline copper sulfate solution (Fehling's reagent while heating. Methylene blue is used as an indicator. During the reaction, Cu_2O precipitates are formed which are brick red in colour which is denoted as the end point.

Apparatus

Conical flask, Burette, glass pipettes

Reagents

- Fehling A: Dissolve 34.639 g of copper sulphate in distilled water in a 500 ml volumetric flask. Make up to volume and mix thoroughly. Filter through Whatman no 1 filter paper and store.
- Fehling B: Dissolve 173 g of sodium potassium tartarate (Rochelle salts) and 50 g of sodium hydroxide in distilled water. Make the volume 500 ml with distilled water.
- Acetic acid: 10 % (w/v) in distilled water

- Methylene blue solution: 1% (w/v) in distilled water
- Standard lactose solution: : 0.5% (w/v) in distilled water

Procedure

Standardization of Fehling's solution

1. Pipette 5 ml of Fehling's solution A and 5 ml of Fehling's solution Busing two separate pipettes in a 250 ml Erlenmeyer flask. Add some inert boiling chips to prevent bumping.
2. Fill up a burette with the standard lactose solution
3. Heat the content of the flask to boiling over burner or heater and maintain moderate boiling for 2 min.
4. Add 3 to 4 drops of methylene blue indicator without removing from the flame.
5. During boiling condition, titrate the content of the flask against standard lactose solution from the burette until the blue colour disappears and the bright brick-red colour of precipitated Cu_2O appears. (at the end point the Cu_2O suddenly settles down giving a clear supernatant).
6. Note the volume of lactose solution used.
7. After this further titration should be carried out, adding practically the whole of the standard lactose solution volume (one ml less than required as observed in first titration) required before commencing the heating and continuing the titration as before. The titration must be completed within 3 min from the commencement of boiling.
8. Let V, ml be the titre for this experiment.

Determination in Milk

1. Heat the sample of milk near boiling, cool to room temperature and take 25 ml in 250 ml volumetric flask.
2. Dilute with 200 ml distilled water and add 10% acetic acid drop by drop till clear precipitation (about 3.75 ml is required).
3. Make the volume to 250 ml with distilled water, mix and filter. Collect the filtrate and fill clean burette with the filtrate.
4. Take 5 ml each of Fehling's A and B in 250 ml conical flask, add 25 ml distilled water and 2-3 pieces of porcelain.
5. Heat it to boiling on a wire gauge and add about 5 drops of methylene blue and titrate against the filtrate containing lactose till blue colour changes to red, maintain moderate ebullition during titration.
6. Take Fehling's solution in another flask as described above and add to it in

cold almost the whole volume of the filtrate required in primary titration, so that not more than 0.5-1.0 ml of it will be required for completing the titration. Heat the mixture to boiling, maintain a moderate ebullition for 2 min, add 5 drops of methylene blue and complete the titration within total time of 3 min by addition of remaining filtrate from the burette. The end point is blue to brick red precipitate. Repeat the titration to get values within 0.1 ml.

Observation

V1 = volume in ml, of standard lactose solution taken to reduce 10 ml of Fehling's solution

V2 = volume in ml, of prepared milk filtrate taken to reduce 10 ml of Fehling's solution.

Calculation

$$\text{Lactose (\% by weight)} = \frac{V_1}{V_2} \times 5$$

Result

The lactose content of the given sample of milk is _____ %

1.7 Determination of Ash Content in Milk

Introduction

Ash is the inorganic residue remaining after the water and organic matter have been removed by heating in the presence of oxidizing agents, which provides a measure of the total amount of minerals within a food. The most widely used methods are based on the fact that minerals are not destroyed by heating, and that they have a low volatility compared to other food components. Principle involved is that when a known weight of sample is ignited to ash, the weight of ash thus obtained is expressed in terms of percentage.

Apparatus

Silica crucible, Water bath, Muffle furnace, Tong, Desiccator

Procedure

1. Weigh empty silica dish which has been ignited and cooled in efficient desiccator.
2. Weigh accurately about 10 g of well mixed sample into the dish.

3. Evaporate the milk in water bath by heating till black residue results.
4. Avoid spurting of milk during heating.
5. Transfer the dish in muffle furnace and incinerate at 550±5°C for 5 h.
6. Cool the dish in desiccator and weigh.
7. Repeat the process of heating cooling and weighing till the difference between two successive weighing is not more than 0.1 mg.
8. Record the lowest weight.

Observation

W2= Weight of crucible after ashing

W1= Weight of empty crucible

W= Weight of sample

Calculation

$$\text{Ash (\% by weight)} = \frac{(w_2 - w_1)}{w} \times 100$$

Result

The ash content of the given sample of milk is _____ %

2 Analysis of Physical Properties of Milk

2.1 Determination of Specific Gravity of Milk

2.1.1 Volumetric Method by Using Lactometer

Introduction

Lactometers are used for rapid determination of specific gravity. The principle of using lactometer is based on law of floatation which states that when a solid is immersed in a liquid it is subjected to an upward thrust equal to the weight of the liquid displaced by it and acting vertically upward. Lactometers are hydrometers of varying immersion types. They are previously calibrated with fluids of known particular gravity, so that the special gravity of the liquid can be read directly. The lactometer sinks more in liquid of low-specific than the liquid of higher specific gravity.

Apparatus

Lactometer, thermometer, lactometer jar, beaker, glass plate

Procedure

1. Heat the milk sample to 40° C for 5 min.
2. Cool the milk and adjust the temperature as nearly as possible to that at which the lactometer is calibrated (84° F for zeal lactometer)
3. Mix the milk sample properly but at the same time avoid air bubble formation.
4. Pour enough milk into the glass cylinder so that some milk overflows when lactometer is inserted.
5. Insert the lactometer gently to wet stem not more than a short length, about 3 m.m beyond the position of equilibrium.
6. The lactometer should float freely and should not touch to the side of the cylinder.
7. Allow the lactometer to remain steady in the milk and note the reading within about 30 seconds and record the temperature of the milk.

Observation

LR = Lactometer reading at 84°F

Calculation

$$\text{Specific gravity of milk (at 84° F)} = \frac{LR}{1000} + 1$$

Result

The specific gravity of the given sample of milk is _____

2.1.2 Gravimetric Method by Using Pycnometer/Specific Gravity Bottle

Introduction

A substance's density is its mass per unit volume and is usually reflected in grammes per ml. The specific gravity of a substance is the ratio of the mass of the substance of a particular volume and the mass at a specific temperature and pressurc of an equal volume of the standard substance. The standard substance is water at 40°C and the unit density is at that temperature for solids and liquids.

The density of milk is directly determined by weighing and determining volume of milk taken in a pycnometer. The specific gravity of milk is calculated by determining weights at the same pycnometer temperature for a certain milk volume and the same volume of water distilled.

Apparatus

Specific gravity bottle or a pycnometer, analytical balance, thermometer

Procedure

1. Carefully clean and dry (with cold dry air) a specific gravity (S.G.) bottle/ pycnometer. It should not be heated to dryness.
2. Weigh accurately the S.G. bottle/pycnometer with the stopper.
3. Fill the S.G. bottle/pycnometer with distilled water without any air bubble and place the stopper tight, overflowing some water.
4. Wipe off water from outside the bottle and weigh accurately.
5. Remove the water and fill the S.G. bottle/pycnometer with milk without any air bubble and place the stopper tight, overflowing some milk.
6. Wipe off milk from outside the bottle.
7. Weigh accurately the S.G. bottle/pycnometer with milk.
8. Note down the temperature of milk and water (in case there is a constant temperature water bath, keep the milk and water in the bath for about 15 min and fill the S.G. bottle/pycnometer with these liquids and record the temperature of the bath).

Observation

W = Weight of empty S.G. bottle with stopper at specific temperature

W1 = Weight of S.G. bottle with stopper and milk at specific temperature

W2 = Weight of S.G. bottle plus distilled water at specific temperature

Calculations

$$\text{Specific gravity (at specific temperature)} = \frac{(w_1 - w)}{(w_2 - w)}$$

Result

The specific gravity of the given sample of milk is _____

2.2 Determination of Viscosity of Milk Using Ostwald Viscometer

Introduction

When a liquid flow through a capillary the viscosity of the liquid is directly proportional to the taken for a certain volume of liquid to flow out under a certain pressure gradient and is also proportional to the density of the liquid.

A comparison of the viscosity of milk is done to that of a standard liquid, distilled water, according to the above principle, and absolute viscosity of milk can be obtained knowing the viscosity of water.

Apparatus

Ostwald's viscometer (1 to 3 centipoise), Stands with clamps to fix the viscometer, Celsius thermometer, Stop watch, Water bath maintained at 20 ± 0.1° C, Pipette, Hair dryer

Reagents

- Milk of cow and buffalo
- Distilled water
- Sodium hydroxide
- Chromic acid

Procedure

1. Clean the viscometer with the sodium hydroxide solution, tap water, chromic acid, tap water and finally four to five times with distilled water; dry the viscometer with cold dust free air.
2. Draw distilled water over the upper mark and close the rubber tube with finger.
3. Viscometer should be properly aligned during measurement of viscosity
4. Release slowly to allow the water to flow down.
5. Start the stop watch when the meniscus of water just leaves the upper mark and measure the time for the meniscus to reach the lower marking.
6. Repeat the experiment at least 3 times and record the times correct up to 0.5 s in a tabular form.
7. Drain out the water from the viscometer and dry it by drawing dust free air.
8. Repeat the experiment similar way with the milk sample.
9. Record the times for at least 3 observations in the tabular form.
10. Measure the density of the milk sample relative to the distilled water used in the experiment a 20°C.

Observation

tm = time in seconds required to come from the upper mark to lower mark for milk tw = time in seconds required to come from the upper mark to lower mark for water

dm = density of milk with respect to distilled water

ŋw = viscosity of water (from standard table)

Calculation

Absolute viscosity of the milk (20°C) $= \frac{tm}{tw} \times \frac{dm}{1} \times \eta w$

Result

The viscosity of the given sample of milk is _____

2.3 Determination of Surface Tension of Milk Using Stalagmometer

Introduction

Surface tension is a property of the surface of a liquid that allows it to resist an external force. Sphere has the minimum surface area. This is why small droplets of liquid are spherical. Surface tension of milk can be determined by comparing the no. of drops of definite volume and densities of milk with a standard liquid (water) at a fixed temperature.

Apparatus

Stalagmometer, Stand and clamp, Beaker, Thermometer, Rubber tubing with a screw pinch cork

Reagents

- Sodium hydroxide
- Chromic acid mixture
- Ethyl alcohol

Procedure

1. Clean the stalagmometer thoroughly with sodium hydroxide to remove grease if any, then with chromic acid mixer and finally with distilled water. Then rinse it with alcohol and dry it by passing a current of air.
2. Attach a small piece of clean rubber tubing free from dust and grease to upper end of the stalagmometer and fix the stalagmometer in a clamp
3. Allow it to attain the same temperature of water.
4. Fill the stalagmometer with distilled water to the upper mark by sucking using the rubber tubing

5. Close the open end of the tube with the tip of the finger.
6. Then release the open end and start counting the drops, till the water level comes from the upper mark to the lower mark.
7. Repeat the process three times more.
8. Remove the rubber tube from the stalagmometer, clean it by rinsing with alcohol and dry it by passing a current of air.
9. Then attach the rubber tube to stalagmometer and fill it with milk, by sucking on the open end of the rubber tube as before.
10. Repeat the procedure three times as before with milk.
11. The surface tension is affected by the temperature; hence the determination is to be carried out at a constant temperature in thermostat.

Observation

y_1 = surface tension of milk

y_2 = surface tension of water

n_1 = number of drops of water

n_2 = number of drops of milk

ρ_1 = density of water

ρ_2 = density of milk

Calculation

Surface tension of milk (y1) $(y_1) = \dfrac{\rho_2 n_2}{\rho_1 n_2} \times y_2 (^\circ C)$

Result

The surface tension of the given sample of milk is _____

2.4 Determination of Redox Potential of Milk

Introduction

Redox potential (Eh) is a parameter of the state of biological media which indicates the capacity to either gain or lose electrons. During oxidation, electrons are transferred from an electron donor to an electron acceptor, which is reduced. Redox potential of milk can be measured by using a calomel electrode and a platinum electrode and measuring the potential difference so developed by a potentiometer or a pH meter.

Apparatus

Cambridge bench model potentiometer cum pH meter with a standard platinum electrode and calomel electrode, Corning beaker with a suitable stopper of about 50 ml capacity to be used as the cell, Thermometer, Water bath, Stop watch

Reagents

- Test sample (milk)
- Quninhydrone
- Phthalate buffer (pH 4.0, 0.5 M)
- HCl - 0.1 N

Procedure

Use the calomel electrode but connect the platinum electrode to the pH meter in place of the usual glass electrode and follow the instructions for reading the potential in milli volts.

Standardisation

Check the Instrument Against

i) pH 4.0 phthalate buffer into which a little quinhydrone has been stirred. Let this reading be ‘a’ volts.

ii) 0.1 N hydrochloric acid also with quinhydrone added. Let this reading be ‘b’ volts

The standard potential of the calomel electrode containing saturated KCl is 0.248V at 20ºC (0.244 V at 25ºC). Assuming that readings are converted to volts at 20ºC, the following values should be obtained

$a + 0.248 = 0.471$ V

$b + 0.28 = 0.642$ V

O-R potential of Milk

1. Take milk in a beaker with a suitable stopper of about 50 ml capacity used as the cell.
2. Adjust temperature of the milk to 30 ± 0.5ºC by dipping in water bath before the electrode combination is inserted into the milk in the cell (50 ml beaker containing milk).

3. To soak the electrodes thoroughly, the sample is stirred for 5 to 10 sec and immediately after this potentiometer reading is taken.
4. pH of milk is measured at 30 ± 0.5°C using the same instrument and dipping a combination of a glass electrode and reference saturated calomel electrode.
5. Determine the redox potential of the test solution (milk) in the same way as for (i) and (ii) steps under the head standardisation described above.
6. Let this reading be x volt, so that, Eh = x + 0.248. If x is negative reverse the electrode and from this reading Eh = - x + 0.248 volts.
7. If a 0.1M KCl calomel electrode is used, the standard potential is V, both at 20 and 25°C.

Calculation

Assuming that the pH is constant, the emf (Eh) at 25°C of a cell consisting of a platinum electrode dipping into a solution containing both forms of the substance (oxidant/ reductant) and connected to a standard hydrogen electrode is given by

$$Eh = E0 = (0.0591/n)* \log10 [Ox]/[Red]$$

Where,

n	=	number of electron involved in the redox system,
[Ox]	=	molar concentration of the oxidized form
[Red]	=	molar concentration of the reduced form and
E0	=	standard redox potential of the system.

Note, If [Ox]> [Red], then Eh> E0, and if [Ox] = [Red], then Eh = E0. The more positive the potential the greater is the oxidising power.

Result

The redox potential of the milk is ________

2.5 Determination of Electrical Conductivity of Milk

Principle

Since milk contains various kinds of ions such as sodium, potassium and chlorides, it can conduct an electric current which can be measured by using a conductivity meter. Conductivity is measured by applying an alternating electrical current (I) to two electrodes immersed in a solution and measuring the resulting voltage (V). During this process, the cations migrate to the negative electrode, the anions to the positive electrode and the solution acts as an electrical conductor.

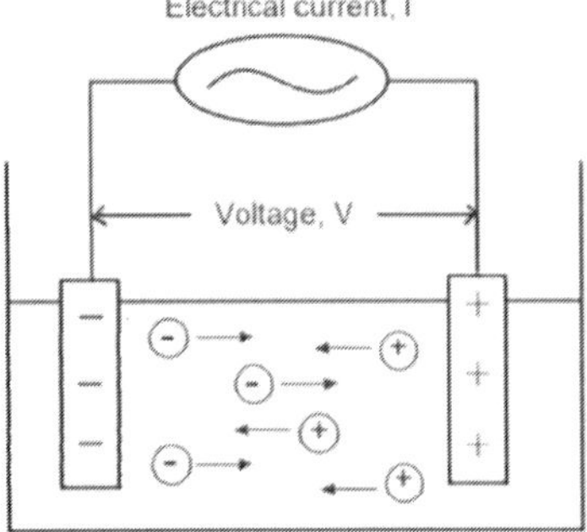

Apparatus

Beaker (250 ml), thermometer, conductivity meter, stabilizer of 230V, 50 c/s

Reagents

- Test sample (milk)
- KCl solution - 1/10 N
- KCl solution - 1/100 N

Procedure

(A) Calibration of a Cell

To calibrate the cell, immerse it in the standard solution (1/10 N and 1/100 N KCl solution) and measure its conductivity, at the same time noting down the solution temperature. Find out the expected conductivity of the solution for this temperature from Table 1.

$$\text{Cell constant} = \frac{\text{Expected conductivity of solution}}{\text{Measured conductivity of the solution}}$$

Table 1: Specific conductivity of standard solutions at a fix temperature

Temperature of solution (^{0}C)	Specific conductivity in mohs	
	1/10 N KCI solution	**1/100 N KCI solution**
15	0.01048	0.001147
16	0.01072	0.001173
17	0.01095	0.001199
18	0.01119	0.001225
19	0.01143	0.001251

20	0.01167	0.001278
21	0.01191	0.001305
22	0.01215	0.001332
23	0.01239	0.001359
24	-	0.001387

(B) Operation of Digital Direct Reading Conductivity Meter

1. When the indicator lamp glows put the CAL/READ switch at CAL position.
2. Adjust the CAL control to full scale reading (199) on the digital panel meter (DPM).
3. Turn the range switch to extreme clockwise position.
4. Immerse the cell in the standard solution, the specific conductivity of which is accurately known at the temperature of solution.
5. Put the CAL/READ switch at READ position and turn the range switch to get the reading in the corresponding range.
6. Now adjust the CELL CONSTANT control so that the DPM shows the right value of the specific conductivity of solution to be measured.
7. Before dipping the cell in any solution, keep the range switch in extreme clockwise direction (position) otherwise in the range the DPM will start blinking due to overflow. Reading at 112 or 113 with flickering or 1 (one) indicate the overflow.
8. To get the specific conductivity of any solution (milk), dip the cell in the solution (after cleaning) the cell by rinsing with distilled water and turn the range switch to get the reading on digital panel meter.

Observation

Electrical conductivity as displayed on the conductivity meter =_____

Result

The given sample of milk has an electrical conductivity of_____.

2.6 Determination of Freezing Point of Milk

Introduction

A solution's freezing point is often less than that of pure solvent, the degree of lowering of solution's freezing point is known as the solution's freezing point depression which is proportional to the solute's concentration.

The freezing point of milk is primarily due to lactose and soluble material salts, and the milk's physical properties are reasonably consistent. The addition of water to the milk lowers the concentration of soluble milk constituents and thereby decreases the freezing point of milk sample added with water, i.e. between the freezing point of pure milk and pure water. The quantity of added water can thus be calculated by calculating the depression of the freezing point of milk sample.

Apparatus

1. Hortvet cryoscope control thermometer (20°C to 30°C, graduation accuracy of 0.1).
2. Standard thermometer (1 to 2°C, graduation of 0.01°C). This is a differential thermometer of Beckman type, specially designed for milk testing.

Reagents

1. Concentrated sulphuric acid
2. Diethyl ether
3. Ethyl alcohol
4. Fresh distilled water
5. Sucrose solution A : 7 g pure sucrose dissolved in distilled water and made up to 100 ml at 20°C.
6. Sucrose solution B: 10 g pure sucrose dissolved in distilled water made up to 100 ml at 20°C.

Procedure

Two standard methods (hortvet and thermistor cryscope) are used for freezing point determination. Both gave same results. The hortvet method is described as below:

i) Hortvet Method

- The temperature difference, in this method, is measured by hortvet mercury-in-glass thermometer.
- Di-ethyl ether is used for cooling of milk,
- But now a day, use of di- ethyl ether is replaced by mechanical refrigeration using 25% (v/v) ethylene glycol for laboratory safety reasons.
- The internal freezing point tubes and fittings are unchanged.

Calibration of the Hortvet Thermometer

1. Standardize the hortvet thermometer with water and against pure sucrose or salt solutions using the procedure described below for milk.
2. Obtain the mean FPD for each sucrose solution by subtraction from the mean of those for water.
3. The corrections to apply for subsequent determinations can be obtained by linear interpolation from the figures in table 2 below:

Table 2: Freezing point depression of salt and sugar solutions

Sodium chloride, g/l (20°C)	Sucrose, g/100 ml (20°C)	Nominal, FPD m °C
6.859	7.0	422
--	7.5	454
7.818	--	480
--	8.0	487
8.314	--	510
8.480	8.5	520
8.563	--	525
8.646	--	530
--	8.75	537
8.811	--	540
8.977	--	550

4. Alternatively correction can be obtained from a single standard whose freezing point is close to that of the sample

Freezing Point of Water

1. Pour 45 ± 1 ml of distilled water into the freezing tube.
2. Start the stirrer at a steady up-and-down motion of 30 to 60 strokes per minute.
3. Continue stirring until the temperature on the hortvet thermometer indicates a temperature -1.3 ± 0.05°C. At this temperature the water may freezing spontaneously.
4. If water does not freeze, insert the freezing starter tube containing fragment of icc, through the hole in the stopper into the water in the tube. It is advisable to keep the starter tube.
5. Keep stirring until the horvet thermometer indicates a rise in temperature and remove the freezing starter tube.
6. Then stir 3 times as the mercury becomes stationary and the thermometer stem 7 times at or a little above the level of the top of the mercury thread

and estimate the reading to 0.001°C with the assistance of a lens and taking precautions to avoid parallax.

7. After 30 s taking the reading and stir 3 times and tap 7 times and read as before.
8. Repeat again after another 30 s and the higher reading (usually the third). The second and third readings should not differ by more than 0.002 °C.

Freezing Point of Milk

1. Determine the acidity of milk.
2. If the acidity exceeds 0.3% (m/v) as lactic acid, the determination of freezing point depression should be abandoned.
3. Adjust the mechanical circulator/cooler so that the temperature in the cooking bath is maintained at -2.5 to 3.0 °C
4. Add sufficient ethanol to the metal tube (about 25 ml) so that the level of alcohol is about 20 mm above the expected level of the milk sample.
5. Rinse the tube, thermometer and stirrer with a portion of the cooled milk sample.
6. Then pour 45 ± 1 ml of milk cooled to 0 to 5°C into the freezing tube and continue as for water above except seed the milk with ice when the reading on the horvet thermometer has fallen to – 1.65°C.
7. The second and third readings should not differ by more than 0.002 °C.

Observation

Tw = Thermometer reading of water

Tm = Thermometer reading of milk

Calculation

Freezing point depression (FPD) = Tw-Tm

Result

The given sample of milk has a freezing point of __________.

2

Analysis of Fat Rich Dairy Products

1 Analysis of Cream

Sample Preparation

Warm the cream sample to 40 - 50°C if the sample is very viscous. Then pour from one vessel to another 3 to 4 times to mix well. Sample should be tested daily and always within three days after collection since enzymatic deterioration of the cream will cause low tests.

1.1 Determination of Fat Content in Cream

1.1.1 Gerber Method

Introduction

Sulphuric acid is added to cream in order to dissolve the protein and other solids and create heat which liquefy and de-emulsify the fat. Iso-amyl alcohol is added to aid the separation of fat. Upon centrifugation, fat being lighter will be separated on top of solution in the cream butyrometer.

Apparatus

Cream butyrometer (0-70%), butyrometer stand, butyrometer stopper and key, centrifuge machine, weighing balance

Reagents

- Gerber acid – 50 % (Sp. gravity - 1.007-1.812 at 20°C)
- Amyl alcohol - (Sp. gravity - 0.803 to0 .805 at 27°C)

Procedure

1. Transfer 10 ml of Gerber acid to a cream butyrometer.
2. Transfer 5 gm of cream to the butyrometer using a funnel. Add 6 ml of distilled water
3. Add 1 ml of iso-amyl alcohol.

4. Insert the stopper and mix thoroughly.
5. Place the butyrometer in the water bath maintained at 65°C for 3 to 10 minutes.
6. Centrifuge the butyrometer at 1100 rpm for 5 minutes.
7. Again place the butyrometer in the water bath at 65°C for 5 minutes.
8. Read the percentage of fat by adjusting the fat column within the scale of the butyrometer.

Observation

BR = Butyrometer reading

Calculation

Percent Fat = (10 x BR) – 1

Result

The given sample of cream has fat content of _____%

1.1.2 Rose-Gottlieb Method

Introduction

This is a gravimetric method. The sample is treated with ammonia and ethyl alcohol. The former is used to dissolve the protein and ethyl alcohol, to precipitate the proteins by reducing the surface tension of the solution. Fat is extracted with diethyl ether and petroleum ether. Mixed ethers are evaporated and the residue is weighed. This method is considered suitable for reference purposes. It is gravimetric method.

Apparatus

Rose Gottlieb tube, Siphon assembly, desiccator, hot plate, conical flask, weighing balance

Reagents

- Ammonia solution – 35 % (Sp. gravity 0.88)
- Ethyl alcohol – 95-96 % (v/v)
- Diethyl ether (Sp. gravity – 0.720, peroxide free)
- Petroleum ether (Sp. gravity- 0.065)
- Sodium chloride solution – 0.5%

Procedure

1. Weight accurately 1g of cream into 50 ml beaker.
2. Add 9 ml sodium chloride solution slowly to disperse and transfer to the fat extraction apparatus.
3. Add 1 ml of concentrated ammonia solution and mix well.
4. Transfer the materials to a Rose Gottlieb tube.
5. Add 10 ml of ethyl alcohol to the beaker and transfer to the Rose Gottlieb tube (fat extraction tube) and mix well.
6. Add 25 ml of diethyl ether to the fat extraction tube. Close the tube with cork (or stopper) and shake vigorously for 1 minute.
7. Remove the cork, add with 25 ml of petroleum ether, wash the cork and the neck of the tube so that the washing run into the tube. Again, cork the tube and mix vigorously for 1 minute.
8. Allow the tube to stand for 30 minutes.
9. Transfer the ether layer to a previously weighed conical flask.
10. Repeat the extraction of the residue by adding 15 ml of diethyl ether and 15 ml of petroleum ether to the fat extraction tube.
11. Mix thoroughly and keep it aside for 10 minutes.
12. Transfer the ether layer to a conical flask and heat the conical flask in hot plate to remove ether and transfer to an oven maintained at 100°C for 1 hour.
13. Cool in a desiccator and weigh.
14. Repeat heating in an oven, cooling and weighing until successive weighing do not show a loss in weight by more than 0.1 mg.

Observation

W1 = Weight in g of the empty flask

W2 = Weight in g of the contents in flask after drying

W = Weight in g of the sample taken

Calculation

$$\text{Fat percent by weight} = \frac{(w_2 - w_1)}{w} \times 100$$

Result

The given sample of cream has fat content of _____%

1.2 Determination of Titratable Acidity of Cream

Introduction

Determination of lactic acid in cream is a valuable indicator of its quality, particularly if it is frozen and stored, since cream may have been neutralized and pasteurized. The acidity of cream is estimated by titrating against standard alkali using phenolphthalein as indicator and expressed as percentage of lactic acid. As one gram equivalent weight of an alkali neutralizes one gram equivalent weight of an acid, the volume of standard alkali required to neutralize lactic acid of fixed volume of cream gives percent acidity of cream milk according to the reaction.

CH_3-CHOH-COOH + NaOH → CH_3CHOH-COONa + H_2O

Lactic acid   Sodium Hydroxide   Sodium lactate   water

Apparatus

Pipette, burette, porcelain basin, weighing balance

Reagents

Sodium hydroxide solution – 0.1N

Phenolphthalein solution - 0 .5 percent in ethyl alcohol

Rosanilin acetate solution

Procedure

1. Weight 10 g cream in two white porcelain basin of approximately 60 ml capacity.
2. Add 10 ml of water to both and stir to disperse the cream.
3. Prepare a colour control in one diluted sample by adding and stirring 2 ml dilute rosaniline acetate solution.
4. In another diluted sample, mix 1 ml phenolphthalein solution
5. Add standard sodium hydroxide (0.1N) solution drop wise as soon as possible until the colour matches the pink colour of the control.
6. The titration should complete within 20 sec.

Observation

V = volume in ml of standard sodium hydroxide required for titration

N = normality of the standard sodium hydroxide solution used

W = mass in g of the sample taken for the test

Calculation

$$\text{Titratable acidity (as \% lactic acid)} = \frac{9NV}{W}$$

Result

The titratable acidity of the cream is ________ % lactic acid

1.3 Detection of Starch in Cream

Introduction

In order to avoid detection of the added water, the starch, cereal meals or arrowroot is generally added to the milk density. Starch is also used as an adulterant and if high amounts of starch are added, this can cause diarrhoea due to the effects of undigested starch in colon.

Apparatus

Test tube, water bath

Reagent

Iodine solution

Procedure

1. Transfer 5 ml of cream in a test tube.
2. Add 5 ml of hot distilled water and mix well.
3. Boil the sample and cool under tap water.
4. Add few drops of iodine solution.
5. Mix thoroughly and observe the colour.

Observation

Appearance of blue colour showed presence of starch in cream.

Result

If blue colour is there: ______

If no blue colour is there: _______

1.4 Detection of Gelatin in Cream

Introduction

Gelatin is mainly used as a thickening agent in dairy products. However, in milk and cream its use is disapproved because it causes an inferior quality of the product. Detection of gelatin in cream generally performed using Stokes method.

Apparatus

Test tube, filter paper

Reagent

- Stokes solution: Dissolve mercury in twice its weight of concentrated nitric acid
- Picric acid

Procedure

1. Transfer 10 ml of cream in a test tube.
2. Add 10 ml of distilled water and mix well.
3. Add 20 ml of stokes solution (dissolve mercury in twice its weight of concentrated nitric acid) and mix gently.
4. Set it aside for 10 minutes and filter.
5. Take 5 ml of filtrate in a test tube.
6. Add 5 ml of picric acid solution.

Observation

Production of yellow precipitate with picric acid indicates the presence of gelatin in cream.

Result

If yellow colour is there then gelatin is ______

If yellow colour is not there then gelatin is ______

2 Analysis of Butter

2.1 Determination of Moisture Content in Butter

Principle

Moisture content of butter is estimated by heating a butter sample in a hot air oven. A known quantity of butter is dried at 102±2°C and the remaining mass is weighed to determinc the loss of weight

Apparatus

Hot air oven, flat bottom aluminium dish (7-8 cm diameter and 2.5 cm deep), glass rod, desiccator, boiling water bath, weighing balance

Procedure

1. Weight accurately 5 g of the butter sample into a previously weighed clean and dry aluminium dish with a glass rod.
2. Place the dish on a boiling water bath for at least 30 minutes, stirring at frequent interval until no moisture can be seen at the bottom of the dish.
3. Wipe the bottom of the dish and transfer it to the oven maintained at 100 ± 1°C and keep it for 30 minutes.
4. Allow the dish to cool in desiccator.
5. Weigh the dish.
6. Repeat the process of heating, cooling and weighing until difference between two successive weights does not exceed 0.1 mg.

Observation

W1 = weight in g of the dish with the sample before heating

W2 = weight in g of the dish with sample after heating

W = weight in g of the empty dish

Calculation

$$\text{Moisture (\% by weight)} = \frac{100(W_1 - W_2)}{W_1 - W}$$

Result

Moisture content of butter = ________ %

2.2 Determination of Curd Content in Butter

Introduction

Curd content in butter is the total solid content present in it. Heating causes separation of that total solid content as curd particles. Petroleum ether is used to remove the fat portion and the remaining residue is dried for determination of curd content. In case of table butter, the salt content can be determined separately by calculating the curd content.

Apparatus

Gooch crucible, Hot air oven, Desiccator, Glass funnels, Filter paper, Moisture dish

Reagents

Petroleum ether – (Sp. gravity – 0.064, boiling point 40 – 60°C)

Procedure

1. Prepare an asbestos mat in a Gooch crucible or sintered funnel, dry in an oven maintained at 100±1°C, cool in a desiccator and weigh, Alternatively, dry, cool and weigh ordinary glass funnel with folded 12.5 cm filter paper.
2. Melt the residue in moisture dish
3. Add 25-50 ml of petroleum ether and mix well.
4. Fit the crucible to the filter flask or place the funnel with filter paper on a filter stand.
5. Wet the asbestos mat or the filter paper with petroleum ether and decant the fatty solution from the dish into the asbestos or the filter paper, leaving the sediment in the dish.
6. Macerate the sediment twice with 20-25 ml of petroleum ether and decant again the fatty solution into the asbestos or filter paper.
7. Filter the solution and collect the filtrate in a clean, dried tared 250 ml flat bottom flask containing a glass bead.
8. With the aid of a wash bottle containing petroleum ether, wash all the fat and sediment from the dish into the crucible or the filter paper.
9. Finally wash the crucible of the filter paper until free from fat, collecting all the filtrate in the flask.
10. Dry the crucible or filter funnel in the oven (100±1°C) for at least 30 min.

11. Cool in a desiccator and weigh the residue.
12. Repeat the drying, cooling and weighing until the loss of weight between two consecutive weighing does not exceed 0.1 mg.

Observation

W1 = weight in g of the filter paper with residue

W2 = weight in g of the filter paper alone

W = weight in g of the sample

Calculation

$$\text{Curd and salt (\% by weight)} = \frac{100\,(W_1 - W_2)}{W}$$

Curd percent by weight in obtained by subtracting the value of salt, % by weight.

Result

The curd and salt percent of the given sample = ______ %

2.3 Determination of Fat Content in Butter by Gerber Method

Introduction

To determine the fat content in butter, butter butyrometer (0-80 percent) is used. The Gerber acid dissolves the protein and other milk solids without charring the fat and creates heat which liquefies and de-emulsifies the fat. Iso-amyl alcohol is added to reduce the surface tension and facilitate the separation of fat. On centrifugation the fat is separated on the top of the solution.

Apparatus

Butter butyrometer (0-80 percent), Gerber centrifuge, Water bath, Glass pipette

Reagents

- Gerber acid – 50 % (Sp. gravity - 1.007-1.812 at 20 °C)
- Iso-Amyl alcohol – (Sp. gravity 0.803 to 0.805 at 27°C)

Procedure

1. Weigh accurately 2.5 g of butter into a 100 ml beaker. Add small quantity of dilute Gerber acid and mix well.

2. Transfer the content to a butter butyrometer with the help of funnel.
3. Wash the beaker 6 times with small quantity of dilute Gerber acid to make sure all the butter has been transferred.
4. Add 15 to 20 ml more dilute Gerber acid to bring the level of content to the butyrometer.
5. Add 1 ml of iso-amyl alcohol. Stopper the butyrometer and mix thoroughly and place in a water bath at 65ºC for 5 minutes.
6. Centrifuge at 1100 rpm for 4 minutes, again place in a water bath at 65ºC for 5 minutes.
7. Note the reading of fat percentage from the graduation scale of the butyrometer.

Observation

Fat percent (%) as displayed on the stem of the milk butyrometer =_____

Calculation

$$\text{Percent fat} = \frac{\text{Fat reading} \times 5}{\text{weight of butter sample}}$$

Result

The fat content of the given sample of milk is _____ %

2.4 Determination of Salt Content in Butter

2.4.1 Volhard's Method

Introduction

Salt percent in butter is extracted by using hot distilled water from the dried fat free residue in moisture determination. Addition of excess amount of silver nitrate causes precipitation of chlorides and the remaining silver nitrate is titrated with potassium thiocyanate using ferric ammonium sulphate as an indicator.

Apparatus

Beaker, volumetric flask, conical flask and water bath (60 – 70ºC)

Reagents

- Standard silver nitrate solution: 0.05 N
- Nitric acid (Sp. gravity 1.42, approx - 70%)
- Nitric acid: 5N
- Ferric ammonium sulphate indicator solution
- Standard potassium thiocyanate solution - 0.05 N

Procedure

1. Extract the salt from the residue of curd and salt (obtained in B 2.2) by repeated washing of the Gooch crucible or filter paper with hot water or by placing the crucible or filter paper in a beaker of hot water.
2. Collect the rinsings in a 100 ml measuring flask, passing the solution through a filter paper.
3. Allow to cool to room temperature and make up to volume.
4. Take 25 ml of water extract in a 250 ml conical flask and add an excess (normally 25-30 ml) of 0.05N silver nitrate solution.
5. Acidify with nitric acid, add 2 ml of the indicator solution (ferric ammonium sulphate) and 1 ml nitrobenzene.
6. Mix and determine the excess of silver nitrate by titration with the potassium thiocyanate (0.05N) solution until the appearance of an orange tint which persists for 15 seconds.
7. In the same manner determine the equivalent of 25 ml or the added amount of silver nitrate as thiocyanate using the same volumes of reagents and water.

Observation

N = normality of potassium thiocyanate

A = volume of potassium thiocyanate in the blank titration

B = volume of potassium thiocyanate in the sample titration

W = weight in g of the sample

Calculation

$$\text{Sodium chloride (\% by weight)} = \frac{23.38 \times N(A-B)}{W}$$

Result

Salt content in the given sample is ______ %

2.4.2 Mohr's Method

Introduction

In this method, the butter sample is melted in hot water, and the chlorides present in the mixture are titrated with a solution of silver nitrate using potassium chromate as indicator.

Apparatus

Conical flask, burette, pipette, measuring cylinder

Reagents

- Standard silver nitrate solution - 0.1 N
- Potassium chromate indicator - 5% (w/v)
- Calcium carbonate: Analytical Grade, free from chloride.

Procedure

1. Weigh accurately about 5g of the sample into the 250 ml conical flask.
2. Carefully add 100 ml of boiling distilled water.
3. Allow to stand for 5-10 min with occasional stirring.
4. After cooling to 50-55°C add 2 ml of potassium chromate solution (5%). Mix by swirling.
5. Add about 0.25g of calcium carbonate and again mix by swirling.
6. Titrate at 50-55°C with standard silver nitrate solution while swirling continuously until the brownish colour persists for half a minute.
7. Carry out a blank test with all the reagents in the same quantity except the sample material. The maximum deviation between duplicate determinations should not exceed 0.02% of sodium chloride.

Observation

N = normality of silver nitrate solution

V_1 = volume of silver nitrate in the sample solution

V_2 = volume of silver nitrate in the blank titration

W = weight in g of the sample

Calculation

Sodium chloride (% by weight) $= \frac{5.85N(V_1 - V_2)}{W}$

2.5 Determination of Titratable Acidity in butter

Introduction

The acidity of butter can be estimated by titrating the butter against a standard alkali using phenolphthalein as indicator and expressed as percentage lactic acid.

Apparatus

Conical flask, burette, beaker

Reagents

- Sodium hydroxide solution - 0.1 N
- Phenolphthalein indicator

Procedure

1. Weigh accurately about 10g of butter sample in a dry 250 ml conical flask.
2. Add 90 ml of hot, previously boiled distilled water and shake the contents.
3. Titrate the hot content with 0.1 N sodium hydroxide solution using 1 ml phenolphthalein indicator.

Observation

V = Volume of sodium hydroxide solution used for titration

N = Normality of sodium hydroxide solution

W = Weight in g of the butter sample

Calculation

Titratable Acidity (as lactic acid) percent by Weight $= \frac{9NV}{W}$

3 Analysis of Ghee

Sample preparation: 50°C

Mix the ghee sample in the container, in which it is received, in a cool place until homogeneous. In the event of any separation taking place between mixing and commencement of analysis of moisture, remix the sample.

After the determination of moisture, place the bottle in a water bath maintained at 50ºC till completely melted. Filter through a Whatman No 4 filter paper with the help of a hot water funnel directly into a received bottle. The filtered ghee should be bright and clear.

3.1 Determination of Moisture

Introduction

The method involves drying a sample in an oven and determining moisture content by the weight difference between dry and wet ghee.

Apparatus

Flat bottom aluminium dish, desiccator, hot air oven, weighing balance

Procedure

1. Weigh accurately about 10g of the sample into a moisture dish which has been dried previously, cooled in a desiccator and then weighed.
2. Place the dish in the oven for approximately 1 h at 105±1°C.
3. Remove the dish from the oven, cool in the desiccator and weigh.
4. Repeat this process of heating, cooling and weighing at half hour intervals until the difference between two consecutive weighings does not exceed 1 mg.

Observation

W_1 = weight in g of the dish with ghee before drying

W_2 = weight in g of the dish with ghee after drying

W = weight in g of the empty dish

Calculation

$$\text{Moisture and volatile matter content (\% by wt.)} = \frac{100(W_1 - W_2)}{W_1 - W}$$

3.2 Determination of Butyro Refractometer Reading (Refractive Index) of ghee

Introduction: 40°C

The refractive index of a transparent medium is usually defined as the ratio of the velocity of light in air to that in the medium. One consequence of refraction is to change in direction of a light ray as it enters or leaves the substance.

Measurement of this bending gives a direct measure of refractive index, n, specifically

$$n = \frac{\sin i}{\sin r}$$

where,

sini = angle of incidence

sin r = angle of refraction

The refractive index of ghee may be measured with the help of abbe or immersion refractometer, which gives refractive index or on a butyro-refractomter, which reads on arbitrary scale Butyro-refractometer (BR) reading at constant temperature. The BR reading of ghee is usually taken at 40°C, since the ghee is completely liquefy at 40°C. Ghee ,vegetable oil and animal body fat exhibits a BR reading at 40°C of about 40 - 43 and above 50, (except coconut: 38-39 ; palm oil : 39-40).Adulteration of ghee with vegetable fat substantially increases in BR reading.

Apparatus

Abbe Butyro-refractometer, thermostatically controlled water bath, beaker

Procedure

1. Clean both the prism of butyro-refractometer with ether and dry
2. Place a few drops of the sample at 40°C on prism. Close prisms firmly.
3. Allow instrument to stand for a few min (2-5) before reading is taken.
4. Maintain temperature at 40°C by circulating by hot water through water bath.
5. Observe the BR reading at 40°C after 2-3 minutes and take the reading.
6. Convert the BR reading into Refractive index with the help of the table provided for the purpose.

The refractive index decreases with rise in temperature and vice versa. If the temperature is not exactly 40°C, X is added to the observed reading for each degree above or subtracted for each degree below 40°C, where

X for Butyro-refractometer = 0.55

X for Abbe refractometer = 0.000365

For conversion of refractive index values into Butyro-refractometer readings and vice versa, is given in the table below.

Table 3: Butyro - refractometer Readings and Indices of Refraction

B.R. Reading	**R.I**	**B.R. Reading**	**R.I**	**B.R. Reading**	**R.I**
35.0	1.4488	40.0	1.4524	45.5	1.4562
35.5	1.4491	40.5	1.4527	46.0	1.4565
36.0	1.4495	41.0	1.4531	46.5	1.4569
36.5	1.4499	41.5	1.4534	47.0	1.4572
37.0	1.4502	42.0	1.4538	47.5	1.4576
37.5	1.4506	42.5	1.4541	48.0	1.4579
38.0	1.4509	43.0	1.4545	48.5	1.4583
38.5	1.4513	43.5	1.4548	49.0	1.4586
39.0	1.4517	44.0	1.4552	49.5	1.4590
39.5	1.4520	44.5	1.4555	50.0	1.4593
		45.0	1.4558		

3.3 Determination of Acidity

Introduction

It is determined by directly titrating the fat in an alcoholic medium against sodium hydroxide solution. It is a relative measure of rancidity as free fatty acids are generally formed during hydrolysis from the glycerides. The value is expressed as percent of oleic acid.

Apparatus

Conical flask, Hot plate

Reagents

- Sodium hydroxide – 0.1 N
- Phenolphthalcin indicator
- Ethyl alcohol

Procedure

1. Weigh 10g of the sample in a 250 ml conical flask.
2. In a second flask bring 50 ml alcohol to boiling point and while still above

70°C neutralize it to phenolphthalein (0.5 ml) with 0.1N sodium hydroxide.

3. Pour the neutralized alcohol on ghee in the flask and mix the contents of the flask.
4. Bring them to boil and when still hot titrate with 0.1N sodium hydroxide, shaking vigorously during the titration.
5. The end point of the titration is reached when the addition of a single drop produces a slight but definite colour change persisting for at least 15 seconds.
6. The result may be expressed in any of the following ways:

Observation

T = volume of 0.1N alkali required for titration, in ml

W = weight in g of sample taken

N = quantity of alkali used expressed as ml of 1N solution

Calculation

- Acid value $= \dfrac{5.61 \times T}{W}$
- Free fatty acids (% oleic acid) $= \dfrac{2.82 \times T}{W}$
- Degree of acidity $= \dfrac{100 \times N}{W}$

3.4 Determination of Soluble and Insoluble Volatile Fatty Acids (Reichert-Meissl, and Polenske Values)

Introduction

The Reichert-Meissl (R.M.) value is the number of ml of 0.1N NaOH aqueous solution required to neutralize the water soluble, steam volatile fatty acids distilled from 5 g of ghee under precise condition specified in the method.

The Polenske value (P.V.) is the number of ml of 0.1N NaOH aqueous solution required to neutralize the water insoluble, steam volatile fatty acids distilled from 5 g of ghee under precise condition specified in the method.

The R.M. value is substantially a measure of the lower fatty acids of ghee like butyric acid (C4:0) and caproic acid (C6:0). Butyric acid contributes about 3/4th and caproic acid 1/4th to the R.M. value, while the P.V. is made up of 1/4th caprylic acid (C8:0) and 3/4th capric acid (C10:0).

In this method ghee is saponified using glycerol-potash diluted with water and acidified, and thereafter steam distilled in glass apparatus at a controlled rate. The condensed and cooled distillate is filtered; the water soluble fatty acids passes through are estimated by titration with alkali to give R.M. value, while the water insoluble acids collected on the filter paper are dissolved out in alcohol and titrated with alkali to give P.V.

Reaction

$$\text{Triglyceride (TAG)} + 3NaOH \rightarrow \text{Glycerol} + \text{Salt of fatty acids}$$

$$R\text{-}COONa + H_2SO_4 \longrightarrow Na_2SO_4 + R\text{-}COOH$$

$$R\text{-}COOH + NaOH \longrightarrow R\text{-}COONa + H_2O$$

Apparatus

Graduated cylinder (100 ml and 25ml), Burette (50ml), Polenske distillation assembly of BIS specification, Glass beads.

Reagents

- Glycerol
- NaOH - 50% m/m
- 25 ml concentrated sulphuric acid diluted to 1lit and adjust until 40ml neutralize 2 ml of 50% NaOH solution, phenolphthalein indicator, ethyl alcohol (95% v/v, neutralized, 0.1N NaOH solution.

Procedure

1. Weigh 5g of ghee into a Polenske flask.
2. Add 20 g of glycerol and 2 ml of 50% sodium hydroxide solution into it.
3. Heat the flask over a naked flame, with continuous mixing, until ghee, including any drops adhering to the upper parts of the flask is saponified, and the liquid becomes perfectly clear
4. Avoid over heating during this saponification.
5. Cover the flask with a watch-glass.
6. Make a blank test without the ghee sample but using the same quantities of reagents and following the same procedure.
7. Measure 93 ml of boiling distilled water, which has been vigorously boiled for 15 min, into a 100 ml cylinder.

8. When the soap is sufficiently cool to permit addition of the water without loss, but before the soap has solidified, add the water and dissolve the soap.
9. If the solution is not clear (indicating incomplete saponification) or is darker than light yellow (indicating over heating), repeat the saponification with a fresh sample of ghee.
10. Add 2 glass beads, followed by 50 ml of the dilute sulphuric acid (1:1), and connect the flask immediately with the distilling apparatus.
11. Heat the flask without boiling its contents, until the insoluble acids are completely melted, then increase the flame and distill 110 ml in between 19 to 21 min.
12. Maintain the flow of water in the condenser such that the temperature of the issuing distillate does not exceed 18-21ºC.
13. When the distillate reaches the 110 ml mark, remove the flame and replace the 110 ml flask by a cylinder of about 25 ml capacities, to catch drainings.
14. Close 110 ml flask with its stopper and without mixing the contents, place it in water at 15ºC for 10 min so as to immerse the 110 ml mark.
15. Remove the flask from water, dry from outside and invert the flask carefully avoiding wetting the stopper with insoluble acids.
16. Mix the distillate by 4-5 double inversions, without violent shaking.
17. Filter through a dry 9 cm open texture filter paper (Whatman No. 4) which fits snugly into the funnel.
18. Reject the first runnings and collect 100 ml in a dry volumetric flask; cork the flask and retain the filtrate for titration.
19. Detach the still head and wash the condenser with three successive 15 ml portion of cold distilled water, passing each washing separately through the cylinder, the 110 ml flask, the filter and the funnel, nearly filling the paper each time and draining each washing before filtering the next. Discard the washings.
20. Dissolve the insoluble acids be three similar washings of the condenser, the cylinder and the filter with 15 ml of neutralized ethanol,
21. Collect the solution in the 110 ml flask and draining the ethanol after each washing. Cork the flask and retain the solution for titration.

Reichert-Meissl or Soluble Volatile Acid Value

- Pour 100 ml of the filtrate containing the soluble volatile acids into a titration flask,
- Add 0.1 ml of phenolphthalein indicator and titrate with the sodium hydroxide solution until the liquid becomes faint pink.

Polenske or Insoluble Volatile Acid Value

- Titrate the alcoholic solution of the insoluble volatile acids after addition of 0.25 ml of phenolphthalein indicator with the 0.1N sodium hydroxide solution until the solution becomes faint pink.

Observation

T1 = volume in ml of 0.1N sodium hydroxide solution used for filtrate of sample

T2 = volume in ml of 0.1N sodium hydroxide solution used for blank of above

T3 = volume in ml of 0.1N sodium hydroxide solution used for alcoholic solution

T4 = volume in ml of 0.1N sodium hydroxide solution used for blank of above

Calculation

Reichert-Meissl value = 1.10 (T1 - T2)

Polenske value = T3 - T4

3.5 Determination of Peroxide Value in Ghee

Introduction

The acceptability of ghee largely depends on the extent to which the oxidative deterioration has occurred. It is generally considered that the first product formed by oxidation of an oil or fat is hydroperoxide. The peroxide further decomposes to secondary oxidation product i.e. aldehyde and ketone which imparts off flavour in ghee. The usual method of assessment of oxidation in ghee is by determination peroxide value (PV) which is reported in units of milliequivalents of peroxide oxygen per kg fat or ml of 0.002N sodium thiosulphate per gm of ghee. The most common method for PV determination is based on iodometric titration. Two methods are recommended for determination PV of ghee i.e. iodometric method and oxygen absorption method.

Hydroperoxides are the first detectable product of autooxidation and are sufficiently stable to keep accumulating for some time. Hydroperoxides are oxidizing agent and they liberate iodine from KI and the liberated iodine can be estimated by titrating against standard sodium thiosulphate using starch as indicator. The liberated iodine is directly proportional to PV of ghee.

Reaction

$$R\text{-}\underset{\mid \atop OOH}{CH}\text{-}R' + CH_3COOK + 2KI \longrightarrow \underset{\mid \atop OH}{R}\text{-}CH\text{-}R' + 2CH_3COOK + H_2O + I_2$$

$$2Na_2S_2O_3 + I_2 + \text{Starch} \longrightarrow 2NaI + Na_2S_4O_6$$

Apparatus

Glass stoppered Erlenmeyer flask, Burette, Water bath

Reagent

- Acetic acid-chloroform mixture (3:2, v/v)
- Freshly made saturated potassium iodide solution
- Sodium thiosulphate solution - 0.1 or 0.01N (made in boiled and cooled distilled water).

Procedure

1. Weight accurately 5 gm ghee in glass stoppered Erlenmeyer flask.
2. Add 30 ml Acetic acid- chloroform mixture and mix it to dissolve.
3. Add 0.5ml saturated solution of potassium iodide, let stand 1min with occasional stirring and add 30 ml distilled water.
4. Titration against 0.1/0.01N sodium thiosulphate solution until yellow colour almost is gone.
5. Add 0.5ml starch indicator then again titration against 0.1/0.01N sodium thiosulphate solution with vigorously shaking until just blue colour just disappeared.
6. Conduct blank determination.

Observation

S = ml of sodium thiosulphate solution

N = normality of sodium thiosulphate solution

Calculation

$$PV\left(\frac{\text{millieqvt peroxide}}{\text{kg oil}}\right) = \frac{\text{SxNx1000}}{\text{weight of sample}}$$

Interpretation

Peroxide value (PV)	Grade
Below 1.5	Very good
1.6 to 2.0	Good
2.1 to 2.5	Fair
2.6 to 3.5	Poor
3.6 to 4.0	Not acceptable

3.6 Determination of Saponification Value

Introduction

Saponification value is defined by the number of mg of potassium hydroxide required to saponify 1 g of fat. The fat sample is saponified by refluxing with alcoholic potassium hydroxide solution. The excess potassium hydroxide solution was then titrated against standard hydrochloric acid solution to know the required amount of alkali needed for saponification of the fat sample.

Apparatus

Round bottom conical flask, Reflux condenser, water bath, burette, glass beads

Reagents

- Alcoholic potassium hydroxide solution – 0.5N
- Hydrochloric acid – 0.5 N
- Phenolphthalein indicator

Procedure

1. Weigh accurately 2.0±0.01g of ghee into a flask.
2. Add 25 ml, accurately measured, alcoholic potassium hydroxide solution (0.5N).
3. Add 1-2 glass beads and boil continuously under reflux condenser for one hour in a boiling water bath, swirling the contents of the flask at frequent intervals.
4. Determine the excess of alkali while the solution is still hot by titration with 0.5N hydrochloric acid, using 0.5 ml of phenolphthalein indicator.

Observation

T_2 = volume in ml of 0.5N acid required for blank

T_1 = volume in ml of 0.5N acid required for the sample

W = weight in g of the sample taken

Calculation

$$\text{Saponification value} = \frac{28.5(T_2 - T_1)}{W}$$

3.7 Determination of Iodine Value

Introduction

Iodine value is defined as the number of grams of iodine absorbed by 100 g of fat using Wij's reagent. The fat sample mixed with carbon tetra chloride is treated with a known amount of excess Wij's solution (iodine mono chloride solution in glacial acetic acid). The excess iodine monochloride is treated with potassium iodide and the liberated iodine is estimated by titrating with sodium thiosulphate solution. Iodine value generally measures the degree of unsaturation in a fat sample

Apparatus

Conical flask, burette

Reagents

- Carbon tetra chloride
- Wij Reagent: Mix 8 gm of iodine trichloride in 350 ml of acetic acid. Separately dissolve 9 gm of iodine in further 450 ml of acetic acid and gently heat. Mix gradually the solution of iodine in acetic acid to the solution of iodine trichloride in acetic acid until the colour has changed to reddish brown. Add 50 ml of more iodine solution and dilute the mixture with acetic acid till 10 ml of mixture is equivalent to 20 ml of standard sodium thiosulphate solution when halogen content is estimated by titration in presence of excess potassium iodide and water. Heat the solution to 100°C for 20 min and cool.potassium iodide solution – 10%
- Sodium thiosulphate solution – 0.1 N
- Starch solution: 1% (w/v) in distilled water

Procedure

1. Weigh accurately 0.40-0.45g of the clear ghee sample in a clean dried conical flask.
2. Dissolve the fat in 15 ml carbon tetrachloride and add by means of a burette exactly 25 ml of the Wij's reagent.
3. Close the flask with its stopper, mix carefully and leave it standing for 1 h in the dark.
4. Add 20 ml potassium iodide solution (10%) and approximately 150 ml of distilled water and mix.
5. Titrate with 0.1N sodium thiosulphate solution (use 2 ml starch solution as indicator), swirling the liquid constantly.
6. Add the starch solution shortly before the end of the titration and shake the contents vigorously.
7. Carry out a blank test using the same quantities of reagents.

Observation

B = volume in ml of standard sodium thiosulphate solution required for the blank

S = volume in ml of standard sodium thiosulphate solution required for the sample

N = normality of the standard sodium thiosulphate solution

W = weight in g of the sample taken in the test

Calculation

$$\text{Iodine value} = \frac{12.05(B - S)}{W}$$

3.8 Detection of Vegetable Fat in Ghee by Phytosterol Acetate Test

Introduction

In this test adulteration of ghee with vegetable oils can be detected. Ghee contains cholesterol and all vegetable oils contain sterols of other types which are together termed as phytosterols. This test only detects the presence of phytosterol, not detect the presence of animal body fat, such as tallow and lard, in ghee, since these contain cholesterol. The sterols are obtained from ghee by crystallization from ethanol of the isolated unsponifiable matter. Then using digitinin these sterols are precipitated from ethanol solution as digitonides. On boiling with acetic anhydride the corresponding acetates are obtained, and their melting points are determined. If the observed melting point of a test sample higher than the normal melting point of sterols (117°C), phytosterol from vegetable fat is assumed to be present and test is positive.

Apparatus

Conical flask, refluxing condenser, water bath, filter paper

Reagents

- Potassium hydroxide solution – 50%
- Ethanol – 96%
- Alcoholic digitonin solution – 1%
- Diethyl ether
- Acetic anhydride

Procedure

1. Weigh about 15g of ghee in a conical flask of 250 ml capacity and add 10 ml of potassium hydroxide solution (about 50% KOH solution) and 20 ml of 96% ethanol.
2. Reflux in boiling water bath for about 30 min to carry out saponification.
3. Add 60 ml of water and then 180 ml of 96% ethanol and raise the temperature to about 40°
4. Add 30 ml of alcoholic digitonin solution (1% digitonin in 96% ethanol), shake and allow cooling.
5. Place the flask in refrigerator at about 5°C for about 12 h.
6. Collect the precipitates of sterol digitonide by filtering through Whatman No. 1 filter paper.
7. Wash the precipitates with water at 5°C, then once with 50 ml of diethyl ether.
8. Dry the precipitates on filter paper in oven at 102°C for 10-15 min.
9. The precipitate represents the sterol digitonides.
10. Weigh about 100 ml of sterol digitonide in a test tube and add 1 ml of acetic anhydride.
11. Heat the tube in a glycerol bath at 145°C until the precipitate has dissolved and continue heating for 2 min.

12. Allow to cool at 80°C.
13. Add 4 ml of 96% ethanol, mix, heat and filter through medium speed filter paper.
14. Heat the filtrate in test tube and bring to boiling and add drop by drop 1-1.5 ml of water until the sterol acetate is just about to precipitate.
15. Add a few drops of 96% ethanol to dissolve again any precipitated sterol acetate. Allow to cool in air for 2 h and finally in ice-water for 30 min. Filter the crystallized sterol acetates and rinse them with 80% ethanol.
16. Re dissolve the crystals by heating in 1 ml of 96% ethanol and allow to cool in air for 15 min and then in ice-water bath.
17. Filter the crystallized sterol acetates and repeat re dissolving, crystallization and filtration to obtain third, fourth and fifth recrystallization.
18. Dry the crystals on paper first at 30°C and then at 120°C for 10-15 min.
19. Grind the crystals in a small agate mortar and fill a melting point tube to a height of about 3 mm.
20. Determine the melting point in the melting point apparatus raising the temperature slowly at a rate of 0.5 degree per min.
21. Take the reading on the thermometer at the moment that the last crystal grain has just disappeared at the melting point.
22. The crystals of sterol acetate may also be observed under the microscope for their typical size and shape.

3.9 Detection of Presence of Sesame Oil (Baudouin Test)

Introduction

The development of permanent pink colour in a sample of ghee sample with furfural solution in the presence of HCl, indicates the presence of sesame oil and the test is known as Baudouin test. According to Govt. of India, Ministry of Agriculture Notification (1962), the edible hardened oil (Vanaspati) shall contain raw or refined sesame (Til) not less than 5% by weight i.e. addition of sesame oil is a must for manufacturer of hydrogenated edible fat. This has been made compulsory because of the specific chromogenic constituent of sesame oil can be easily detected, and hence adulteration of ghee with hydrogenated fats can be easily established.

The unsaponifiable matter of sesame oil contains two chromogenic constituent namely sesamolin and sesamol which are not found in other fats. The sesamol on condensation with furfural produces the pink colour in the Baudouin test.

Apparatus

Stoppered measuring cylinder – 25 ml

Reagents

- Hydrochloric acid (fuming): Sp. gravity - 1.19
- Furfural solution - 2% solution of furfural, in ethanol (distilled not earlier than 24h prior to the test from rectified spirit).

Procedure

1. Take 5 ml of the melted ghee in a 25 ml measuring cylinder (or test tube) provided with a glass stopper
2. Add 5 ml of hydrochloric acid and 0.4 ml of furfural solution.
3. Insert the glass stopper and shake vigorously for 2 min.
4. Allow the mixture to separate.
5. The development of pink or red colour in the acid layer indicates the presence of sesame oil.
6. Confirm by adding 5 ml of water and shaking again.
7. If the colour in the acid layer persists, sesame oil is present.
8. If the colour disappears, it is absent.

3.10 Detection of Presence of Cottonseed Oil

3.10.1 Halphen's Test

Introduction

The development of red colour on heating the ghee with a solution of sulphur in carbon di sulphide indicates the presence of cotton seed oil. The test is also given by hempseed oil, Kapokseed oil/oils and fat containing cyclopropinoid fatty acids (such as sterculic and malvelic acid). The fat of animals fed on cottonseed meal or other cotton seed products may give faint positive reaction by this test.

Apparatus

Test tube, Brine water bath (110°C – 115°C)

Reagent

Sulphur solution

Procedure

1. Take 5 ml of the ghee sample in a test tube and add 5 ml of amyl alcohol and mix well.
2. Add 5 ml of 1% solution of sulphur prepared in carbon disulphide.
3. Mix the contents thoroughly and place the tube in saturated sodium chloride bath for 30-35 min.
4. Development of rose red colour indicates the presence of cottonseed oil in ghee.

3.10.2 MBR Test

1. Take about 5 ml of molten ghee sample in a test tube.
2. Add 0.1 ml of 0.1% methylene blue dye prepared in methanol: chloroform (1:1, v/v) mixture.
3. Mix the contents thoroughly and observe for instant disappearance of colour, which indicates the presence of cottonseed oil in the sample.

3

Analysis of Cheese

Sample Preparation

Cheese sample is prepared by passing through 8 mesh sieve three times or grate sample and mix thoroughly or grind to a uniform mass in a glass pestle and mortar. Keep the sample in an airtight container until the time of analysis.

1 Determination of Moisture Content in Cheese

Introduction

The moisture content of cheese is defined as the loss in mass, expressed as a percentage by mass when the product is heated in a hot air oven at 102 ± 2°C to constant mass.

Apparatus

Shallow flat bottom dishes or porcelain dishes, water bath, hot air oven, glass rod,weighing balance

Procedure

1. Weight accurately a dried empty dish with glass rod
2. Weight 3 – 5 g sample in the dish and spread the sample evenly at the bottom of dish with help of glass rod.
3. Transfer the dishes into a hot air oven 102±1 °C for 4 hrs.
4. Remove the dish, cool in desiccators and weight.

Observation

W1 = Weight of empty dish with glass rod

W2 = Weight of empty dish with glass rod and sample

W3 = Weight of empty dish with glass rod and dried residue

Calculation

$$\%Moisture = \frac{W_3 - W_1}{W_2 - W_1} \times 100$$

2 Determination of Fat Content in Cheese

2.1 Mojonnier Method

Introduction

In this method, proteins are digested with concentrated hydrochloric acid. Liberated fat is extracted with alcohol, ethyl ether and petroleum ether. Ethers are evaporated and residue left behind is weighted to calculate the fat content.

Apparatus

Mojonnier fat extraction tube, water bath, hot air oven, etc.

Reagent

- Hydrochloric acid (Sp. gravity - 1.125)
- Ethyl alcohol (95-96% v//v)
- Diethyl ether (Sp. gravity - 0.72)
- Petroleum ether (Sp. gravity – 0.064, boiling point - 40-60°C)

Procedure

1. Weigh accurately 1-2 g of prepared sample (processed Cheese, processed cheese spread and soft cheese) in a 100 ml beaker.
2. Add 10 ml of conc. hydrochloric acid. Heat on a Bunsen burner, stirring continuously with a glass rod, or on a boiling water bath until all solid particles are dissolved.
3. Cool to room temperature and add 10 ml of ethanol.
4. Add 25ml of diethyl ether; shake vigorously for 1min.
5. Add 25ml of petroleum ether; shake vigorously for 1min.
6. Allow to separate ethereal layer or use mojonnier centrifuge at 600rpm ≥ 30 seconds to obtain clear ethereal layer.
7. Decant the upper clear layer.
8. Repeat extraction twice by using 15ml each of diethyl ether and petroleum ether.

9. Evaporate the solvent of dishes on a hot plate.
10. Dry the residual fat in the oven at 100±1°C for 1hour.
11. Transfer the dish in desiccator and observe the weight difference.

Observation

W1 = Weight of sample taken

W2 = Weight of empty dish

W3 = Weight of dish with dry fat residue

Calculation

$$\%Fat = \frac{W_3 - W_2}{W_1} \times 100$$

2.2 Gerber Method

Introduction

The cheese sample is mixed with sulphuric acid and iso-amyl alcohol in a special Gerber tube, permitting dissolution of the protein and release of fat. The tubes are centrifuged and the fat rising into the calibrated part of the tube is measured as a percentage of the fat content of the milk sample. The method is suitable as a routine or screening test. It is an empirical method and reproducible results can be obtained if procedure is followed correctly.

Apparatus

Gerber centrifuge machine, cheese butyrometer (0-40%), butyrometer stopper, water bath.

Reagents

- Gerber acid – 90 % (Sp. gravity - 1.007-1.812 at 20°C)
- Iso amyl alcohol – (Sp. gravity 0.803 to 0 .805 at 27°C)

Procedure

1. Weight accurately 3gm of cheese in cheese butyrometer.
2. Add 10ml of sulphuric acid (90-91%) and 10ml distilled water.
3. Add 1ml iso amyl alcohol.
4. Digest the content by mixing.

5. Adjust the volume in butyrometer by adding water.
6. Centrifugation 1200 rpm for 5min.
7. Place the butyrometer in water bath 65°C for 5 min.
8. Read % fat of the given sample from the butyrometer stem
9. If there are less than 40% fat in cheese than take 1.5gm sample for analysis and multiplied %fat by two.

3 Determination of Salt in Cheese

Introduction

Generally, sodium chloride is added for cheese making. To determine the sodium chloride cheese sample is initially dissolved in warm distilled water. Then the salt is estimated by titrating the cheese solution against standard silver nitrate solution using potassium chromate as indicator until brick red colour persists for at least 30 seconds.

$$AgNO_3 + NaCl \longrightarrow AgCl + NaNO_3$$

$$AgNO_3 + K_2CrO_4 \longrightarrow Ag_2CrO_4 + KNO_3$$

Apparatus

Volumetric flask, burette, conical flask, pipette

Reagents

- Silver nitrate solution – 0.1N
- Potassium chromate solution – 1 %

Procedure

1. Weigh accurately 2 g of cheese sample into a small beaker.
2. Add 10 ml of warm distilled water (50-55°C) and make it paste.
3. Add further 25 ml of warm distilled water and stir well.
4. Transfer the content to a 100 ml volumetric flask. Rinse the beaker with distilled water several times and transfer rinsing to volumetric flask.
5. Make up to volume with warm distilled water and mix thoroughly and filter.
6. Pipette out 25 ml of above filtrate into a conical flask.
7. Add 1 ml of potassium chromate solution and titrate with silver nitrate solution until brick red colour persists.

Observation

V = Volume in ml of silver nitrate solution used for titration

W = Weight in g of cheese sample

Calculation

Salt (sodium chloride) percent by weight $= \frac{v \times 0.585 \times 4}{w}$

4 Determination of pH in cheese

Introduction

pH is the negative logarithm of hydrogen ion concentration and is determined using a pH meter.

Apparatus

pH meter, beaker, mixer etc.

Reagent

- Buffer pH 4.0 and 7.0

Procedure

1. Calibrate the pH meter using standard pH buffer 4.0 and 7.0
2. Weight accurately 30 g of grated cheese.
3. Add 75ml distilled water.
4. Mix thoroughly.
5. Measure the pH of solution in pH meter.

4

Analysis of Ice Cream

Sample Preparation

a) **Plain products:** Sample should be softened at room temperature. Since the melted fat tends to separate and rise to the surface, it is not advisable to soften the sample by heating on a water bath or oven or flame. Mix thoroughly by stirring with spoon or eggbeater or by pouring back and forth between beakers.

b) **Ice-cream containing Fruits and Nuts:** Fill one third part of the cup with ice cream sample. Melt the product at room temperature and mix until insoluble particles are finely divided (about 3-5 minutes for fruit pieces and upto 7 minutes for nut pieces). Otherwise, the product may be ground in a porcelain or glass pestle and mortar. Don't let temperature exceed 25°C at any time during softening and mixing steps. If fat separation occurs, discard sample otherwise immediately pour mixture into wide jar and cap tightly. If allowed to stand, shake vigorously before taking sample for analysis.

c) In case of ice cream, where the chocolate or similar covering portions forms a separate layer, it should be removed and only the ice cream portion should be taken for analysis.

1 Determination of Overrun of Ice Cream

Introduction

Overrun is usually defined as the volume of ice cream obtained in excess of the volume of the mix. It is usually expressed as percentage of overrun. This increased volume is composed mainly of air incorporated during the freezing process. The amount of air that should be incorporated depends upon the composition of the mix and the way it is processed, and is regulated so as to give the percentage of over run or yield that will give proper body, texture and palatability necessary to good quality ice cream.

Apparatus

Analytical balance, metal cup/ice cream empty cup

Procedure

1. Weight the empty cup accurately.
2. Fill the cup up to marked with ice cream mix and weight it accurately.
3. During the manufacture, fill the cup to the marked level with ice cream and weight it accurately.
4. Repeat the weighing at different interval of time during freezing process.

Observation

W1 = Weight of empty cup

W2 = Weight of cup with ice cream mix

W3 = Weight of cup with ice cream

Calculation

$$\%\text{Over run} = \frac{(w_2 - w_1) - (w_3 - w_1)}{w_3 - w_1} \times 100$$

2 Determination of Total Solids in Ice Cream

Introduction

A known quantity of the sample is dried with sand at 100°C. Sand is used to avoid the burning of the sample during heating. The loss in weight is moisture due to evaporation.

Apparatus

Aluminium dish, glass rod, weighing balance, hot air oven, beaker, sand

Preparation of sand: It should be passed through 500- micron IS sieve and retained on 180-micron IS sieve. It should be prepared by digestion with concentrated hydrochloric acid, followed by thorough washing with water to free from chlorides. It should be dried and ignited to dull red heat.

Procedure

1. Weight accurately a dry empty dish.
2. Weight about 3gm of well mixed sample of ice cream accurately.
3. Place the dishes on boiling water bath till apparently dry.

4. Transfer the dishes to hot air oven at 102±2°C for 4 hrs.
5. Transfer the dishes to a desiccator and allow to cool.
6. Weight the dishes accurately.
7. Repeat, if necessary, until the loss of weight is not more than 0.5mg

Observation

W_1 = Weight of dish

W_2 = Weight of dish with ice cream

W_3 = Weight of dish with dried ice cream

Calculation

$$\%\text{Total solids} = \frac{W_3 - W_1}{W_2 - W_1} \times 100$$

3 Determination of Acidity in Ice Cream

Introduction

The acidity of the ice-cream shall be determined before the addition of colouring matter.

Apparatus

Conical flask, burette

Reagents

- NaOH solution - 0.1N
- Phenolphthalein indicator - 1%

Procedure

1. Weight accurately 20 gm prepared sample in a conical flask.
2. Add 50 ml boiled and cooled distilled water.
3. Ml phenolphthalein indicator.
4. Titration against 0.1N NaOH till light pink appeared.

Observation

V = Volume of 0.1 N NaOH

N = Normality of NaOH

W = Weight of sample

Calculation

$$\text{Titratable acidity (\% of lactic acid)} = \frac{9xVxN}{W}$$

4 Determination of Fat in Ice Cream

4.1 Mojonnier Method

Introduction

In this method, proteins are digested with concentrated hydrochloric acid. Liberated fat is extracted with alcohol, ethyl ether and petroleum ether. Ethers are evaporated and residue left behind is weighted to calculate the fat content

Apparatus

Fat extraction flask or Mojonnier tube, hot plate, oven etc.

Reagent

- Ethyl alcohol - 95-96% (v/v)
- Diethyl ether (Sp. gravity - 0.720)
- Petroleum ether (Sp. gravity 0.065, boiling point-40-60°C)

Procedure

1. Accurately weigh 4-5 g of the thoroughly mixed sample directly into fat extraction flask or Mojonnier tube
2. Dilute with water to approximately 10 ml, working sample into lower chamber and mix by shaking.
3. Add 2 ml ammonia; mix thoroughly, heat in water bath for 20 min at 60°C with occasional shaking, cool the content.
4. Add 10ml ethyl alcohol; shake for 15 seconds.
5. Add 25ml of diethyl ether; shake vigorously for 1min.
6. Add 25ml of petroleum ether; shake vigorously for 1min.
7. Allow to separate ethereal layer or use Mojonnier centrifuge at 600rpm ≥ 30 secconds to obtain clear ethereal layer.
8. Decant supernatant layer.
9. Repeat extraction twice by using 15ml each of diethyl ether and petroleum ether.
10. Evaporate the solvent of dishes on a hot plate.

11. Dry the residual fat in the oven at 100 ±1 °C for 1hour.
12. Transfer the dish in desiccator and observe the weight.

Observation

W_1 = Weight of sample taken

W_2 = Weight of empty dish

W_3 = Weight of dish with dry fat residue

Calculation

$$\%Fat = \frac{W_3 - W_2}{W_1} \times 100$$

4.2 Gerber Method

Introduction

The ice cream sample is mixed with sulphuric acid and iso-amyl alcohol in an ice cream butyrometer permitting dissolution of the protein and release of fat. The tubes are centrifuged and the fat rising into the calibrated part of the tube is measured as a percentage of the fat content of the milk sample. The method is suitable as a routine or screening test. It is an empirical method and reproducible results can be obtained if procedure is followed correctly.

Apparatus

Centrifuge machine, ice cream butyrometer, stopper, water bath

Reagent

- Gerber acid – 90% (Sp. Gravity - 1.78)
- Iso amyl alcohol - (Sp. gravity - 0.803 to 0 .805 at 27°C)

Procedure

1. Weight carefully 5gm of melted sample into butyrometer.
2. Add 6 ml of hot water for dilution and washing.
3. Pour 10 ml of sulphuric acid into butyrometer.
4. Add 1 ml iso amyl alcohol.
5. Shake and centrifuge at 1100-1200 rpm for 5min.
6. Keep in water bath at 60°C for 5min and read the % fat.

5 Determination of Sugar Content in Ice Cream

Introduction

This determination includes three main stages: (a) quantitative partitioning of sucrose into aqueous phase from the product, (b) determination of original reducing sugar content and determination of total reducing sugar content after inversion.

The quantitative isolation of sugar is accomplished by clarifying a portion of the product by combination of lead acetate and alumina cream (for ice cream and dried ice cream mix). The precipitated material is washed several times to collect the traces of carbohydrates present in it.

Sucrose is basically a non-reducing sugar but its hydrolysis (inversion) product is reducing. Therefore, if aqueous phase is used as such to reduce the cupric (Cu^{++}) to cuprous ion (Cu^{+}), it estimates and after inversion estimates total reducing sugar. Sucrose content is equal to total reducing sugar less original reducing sugar, provided that no non-reducing sugar other than sucrose is present in the sample.

When Fehling's A and B are mixed, a complex between copper hydroxide and sodium potassium tartrate is formed. Hence precipitation of copper hydroxide is prevented. When Fehling solution heated in the presence of reducing sugar the complex cupric ion is reduced to cuprous ion and brick red precipitate of cuprous oxide are formed.

Reagent

- Sodium hydroxide solution - 0.1N
- Stock solution of invert sugar
- Standard solution of invert sugar
- Methylene blue indicator solution
- Fehling's solution- Prepared by mixing immediately before use, equal volumes of Fehling's solution A and Fehling's solution B.
- Alumina Cream
- Neutral lead Acetate Solution
- Concentrated Hydrochloric acid solution - (Sp. Gravity - 1.18)
- Concentrated ammonia solution - (Sp. gravity - 0.88)
- Saturated solution of sodium oxalate

Standardization of Fehling's Solution

1. Pipette accurately 5 ml of each solution Fehling's 'A' and 'B' into a 250 ml Erlenmeyer flask.
2. Add (about 20ml) from burette a volume of less than 1ml than the expected volume of diluted invert sugar solution that will completely reduce Fehling solution and sufficient water to bring the volume 75ml at the commencement of boiling.
3. Then add 1ml of methylene blue solution, continue the heating and addition of invert sugar solution drop wise until the titration is complete, which is shown by the reduction of the dye (The end point of the titration is blue to red do not stir the flask, mixing is achieved by continuous boiling).
4. Note down the volume of invert sugar solution required to reduce all the copper.
5. Repeat the titration, adding before heating almost all of the dextrose solution needed to reduce all the copper so that not more than 0.5-1 ml. is required later to complete the titration. Heat the mixture to boiling and boil gently for 2 minutes, lowering the flame sufficiently to prevent bumping.
6. Complete the titration within a total boiling time of about 3 minutes, by small additions of invert sugar solution to complete de-colorization of the indicator.
7. Note down the volume of dextrose solution required to reduce all the copper and calculate the factor for the Fehling's solution.
8. Factor = Titer value x mg.invert sugar solution in 1.0 ml

Inversion procedure

1. Weigh accurately about 10g of the well - mixed sample and transfer to 250 ml conical flask and dilute to 150 ml with water.
2. Add neutral lead acetate drop by drop, mixing by rotating the flask until no further precipitate is formed.
3. Add one drop of alumina cream and allow to stand for few minutes with occasional rotation to ensure complete precipitation of protein.
4. Add just sufficient sodium oxalate solution to precipitate excess lead.
5. Filter through whatman no.1 paper into 250ml volumetric flask.
6. Wash the precipitate and the paper thoroughly with hot water collecting washing in the volumetric flask.
7. Cool the flask and content and make up to the mark (solution C).

8. Carry out titration against Fehling solution prepared and standardized above.
9. Concentration of reducing sugar in the solution should be such that the titer value should be within 15-50ml. If it is not, then adjust dilution of the solution.

Determination of reducing sugar after inversion

Procedure

1. Transfer 50 ml filtrate (solution C) to 250 ml volumetric flask.
2. Add 25ml of water and 10ml 6.34N HCl.
3. Partially immerse the flask in water bath adjusted to 70°C and when temperature of content reaches to 67°C (which should take 2.5 - 2.75min), allow further 5min heating by that time temperature should have reached about 69.5°C.
4. Remove the flask and cool immediately by immersing in water bath at 20°C.
5. When contents of the flask have nearly reached 20°C, make just neutral to litmus by adding 1N NaOH and dilute to mark with 20°C chilled water.
6. Determine the total reducing sugar by titrating against Fehling solution.

Observation

F = mg invert sugar equivalent to 10 ml Fehling solution

S = burette reading of titration before inversion

T = burette reading of titration after inversion

W = Weight of sample taken for analysis

Calculation

Orignal reducing sugar (ORS), as invert sugar % by mass $= \dfrac{25xF}{SW}$

Total reducing sugar (TRS) as invert sugar % by mass $= \dfrac{125xF}{SW}$

Sucrose (% by mass) = [TRS%-ORS%] x 0.95

6 Determination of Protein Content in Ice Cream by Kjeldahl Method

Introduction

The protein content is determined from the organic Nitrogen content by Kjeldahl method. The various nitrogenous compounds are converted into ammonium sulphate by boiling with concentrated sulphuric acid. The ammonium sulphate formed is decomposed with an alkali (NaOH) and the ammonia liberated is absorbed in excess of standard solution of acid and then back titrated with standard alkali. The percent nitrogen obtained is multiplied by a factor to get protein content in ice cream sample.

Apparatus

Digestion block, Digestion tubes, Exhaust manifold, Measuring cylinder, Distillation unit, Conical or Erlenmeyer flask, Burette

Reagents

- Potassium sulfate (K_2SO_4)
- Copper (II) sulfate solution
- Concentrated sulphuric acid
- Sodium hydroxide solution, 50%, m/m (low in nitrogen)
- Indicator solution: Dissolve 0.1 g of methyl red in 95% (v/v) ethanol and dilute to 50 ml with ethanol. Dissolve 0.5 g of bromocresol green in 95% (v/v) ethanol and dilute to 250 ml with ethanol. Mix 1 part of methyl red solution with 5 parts of bromocresol green solution or combine all of both solutions.
- Boric acid solution (H_3BO_3), 4%
- Standard Hydrochloric acid solution: 0.1 ± 0.0005 N.
- Ammonium sulfate [$(NH_4)_2SO_4$] : Minimum assay 99.9%
- Tryptophan ($C_{11}H_{12}N_2O_2$) or Lysine hydrochloride ($C_6H_{15}ClN_2O_2$): Minimum assay 99%
- Sucrose with a nitrogen content of not more than 0.002% (m/m).

Procedure

Digestion of sample

1. Take 6 ± 0.1 g of prepared ice cream sample in a clean and dry digestion tube
2. Add 12 g K_2SO_4, 1.0 ml of the copper sulfate solution in to the tube

3. Add 20 ml of concentrated sulfuric acid.
4. Gently mix the contents of the tube.
5. Set the digestion block at a low initial temperature so as to control foaming (approximately 180 – 230°C).
6. Transfer the tube to the digestion block and place the exhaust manifold which is itself connected to a centrifugal scrubber of similar device in the top of the tube.
7. Digest the test portion for 30 min or until white fumes develop. Then increase the temperature of digestion block to a temperature between 410 – 430°C and continue digestion of the test portion until the digest is clear.
8. After the digest clears (clear with light blue-green colour) continue digestion at a temperature of between 410 – 430°C for at least 1 h. The total digestion time will be between 1.75 – 2.5 h.
9. Remove the tube from the block with the exhaust manifold in place. Allow to cool to room temperature over a period of approximately 25 min.
10. The cooled digest should be liquid or liquid with a few small crystals at the bottom of the tube. Do not leave the undiluted digest in the tube overnight. The undiluted digest may solidify and it will be very difficult to get that back into the solution with water.
11. After the digest is cooled to room temperature, remove the exhaust manifold and carefully add 85 ml of water to each tube. Swirl to mix while ensuring that any crystals which separate out are dissolved. Allow the contents of the tube to cool again to room temperature.

Distillation

1. Turn on the condenser water for the distillation apparatus. Attach the digestion tube containing the diluted digest to the distillation unit. Place a conical flask containing 50 ml of the boric acid solution under the outlet of the condenser, in such a way that the delivery tube is below the surface of the boric acid solution. Adjust the distillation unit to dispense 55 ml of sodium hydroxide solution.
2. Continue with the distillation process until at least 150 ml of distillate has been collected. Remove the conical flask from the distillation unit.

Titration

1. Titrate the boric acid receiving solution with standard hydrochloric acid solution (0.1 N) to the first trace of pink colour. Take the burette reading to at least the nearest 0.05 ml. A lighted stir plate may aid visualization of the end point.

Blank test

1. Simultaneously carry out a blank test by following the procedure as described above taking all the reagents and replacing the milk sample with 0.85 g of sucrose.

Observation

N = Normality of acid

S = Titer value of sample

B = Titer value of blank

W = Weight of sample taken

Calculation

$$\%\text{Nitrogen} = \frac{1.4007 \times (S - B) \times N}{w}$$

% Protein (ice cream/Kulfi) = % Nitrogen x 6.38

% Protein (Frozen dessert) = % Nitrogen x 6.25

5

Analysis of Dried Milk

Sample Preparation

Sample should be taken at six or more points on the top of the surface by means of a tubular trier of sufficient length to extend to the bottom of the barrel. Remove this cores and transfer to clean, dry, air tight container and seal immediately.

Before opening sample for analysis, make homogenous either by shaking or by alternately rolling and inverting container. If lump are present, sift sample through No 20 sieve, rubbing material through sieve and tapping vigorously, if necessary.

Avoid sampling on rainy day or when humidity is high, so as to reduce moisture absorption from air.

1 Determination of Moisture Content by Moisture Balance

Introduction

In this procedure, infra-red radiation is used to accomplish rapid drying. The tester has a sensitive torsion balance which permits rapid determination directly from the balance scale, eliminating the need for computing or calculation to obtain results. This method is useful for the in-process determination of moisture because the results may be obtained in minutes.

Apparatus

Infra-red moisture balance

Procedure

1. Light the scale lamp by means of toggle switch at the right side of the balance.
2. Rotate the scale until 100 percent mark on the graduated scale is at the index line.
3. Bring pointer to the index line by turning the pointer adjusting knob on left side of the unit (This calibrates 'zeros' in the balance).

4. Rotate scale until zero percent is at index line. The instrument is then prepared to receive the sample.
5. Move the toggle switch to the off position.
6. Raise lamp housing and carefully distribute approximately 5 g of milk powder on the sample pan until pointer return to the index line (This amount of material is then corresponds to 100 divisions on the scale).
7. Close lamp housing, move the toggle switch to the on position and rotate the scale until 80 percent line on the graduated scale is lined up with the index line.
8. This keeps the sample pan level and allow for even drying. Set timer for 4 to 41/2 minute.
9. As the sample loses moisture, the pointer will rise above the index line.
10. Then switch off heat lamp after 4 to 5 minutes (or until the pointer remains motionless for 30 seconds to 1 minutes).
11. To determine percent reduction in weight due to loss of moisture, adjust the position of the pointer to line up with index line by rotating the scale.

Observation

Read directly from the scale as percent moisture

2 Determination of Total Solids

It is calculated by subtracting the moisture percentage from 100.

3 Determination of Titratable Acidity

Introduction

The titratable acidity can be measured by titrating milk powder solution against a standard alkali using phenolphthalein as an indicator and is expressed in term of lactic acid.

Apparatus

Conical flask, burette, pipette, glass rod

Reagents

- Sodium hydroxide solution – 0.1N
- Phenolphthalein solution – 0.5 percent in ethyl alcohol, 95 percent by volume

3.1.1 ISI Method

Procedure

1. Weigh accurately about 2g of milk powder and transfer into 250 ml conical flask or beaker.
2. Add 20 ml of hot distilled water. Stir vigorously until a smooth liquid is obtained.
3. Cool, add 1 ml of phenolphthalein indicator.
4. Titrate against 0.1N sodium hydroxide solution. Stir vigorously throughout the titration.
5. Time taken for titration should not exceed 20 seconds.

Observation

V = volume in ml of standard sodium hydroxide used for titration

W = weight in g of milk powder

Calculation

$$\text{Titratable acidity (as \% lactic acid)} = \frac{0.9 \times V}{W}$$

3.2 ADMI method

- Dissolve and disperse 10g of skim milk powder or buttermilk powder or 13g of whole milk powder in 100 ml distilled water by using a mixer.
- Allow the sample to stand for about 1 hour.
- Stir gently, and then pipette out 17.6 ml of reconstituted sample into a beaker.
- Rinse out the same pipette with 17.6 ml of distilled water and add this also to the beaker.
- Add 0.5 ml of phenolphthalein indicator, titrate with 0.1N sodium hydroxide solution, until a faint pink colour persists for 30 seconds.

Observation

V = volume in ml of standard sodium hydroxide used for titratio

Calculation

$$\text{Titratable acidity (as \% lactic acid)} = \frac{V}{20}$$

4 Determination of Solubility Index

Introduction

The solubility index is a measure of the un- dissolved residue of milk powder. It is the volume (ml) of sediment remaining at the bottom of a centrifuge tube after centrifuging a known quantity of milk powder dispersed in water. It is mainly the consequence of milk protein denaturation; the degree of denaturation determines powder solubility to a greater extent.

Apparatus

Weighing balance, Centrifuge with cups to accommodate conical centrifuge tubes.

Procedure

1. Add 13g of whole milk powder and partly skimmed milk powder or 10g of skimmed milk powder to 100 ml of distilled water, at a temperature of 24°C in a mixing jar, and stir for exactly 90 seconds.
2. Allow the sample to stand until the foam has separated sufficiently to permit its complete removal by a spoon.
3. The period of standing after mixing should not exceed 15 min.
4. After removal of foam, mix the sample thoroughly with a spoon for 5 seconds.
5. Fill the centrifuge tube immediately with the reconstituted milk to the 50 ml mark. Centrifuge the tube for 5 min at the required speed.
6. Immediately siphon off the transparent liquid within 5 ml of the surface of the sediment level, taking care not to disturb the sediment layer.
7. Add about 25 ml distilled water at a temperature of 24°C. Invert and shake to mix the contents thoroughly.
8. Re-centrifuge at the required speed for 5 min.
9. Hold the tube in a vertical position and read the milliliters of sediment in the tube to the nearest graduated scale division.
10. Report the solubility index as the milliliters of sediment in the tube.

5 Determination of Solubility Percentage

Introduction

The powder is mixed with water and the total solids of the suspension is determined before and after centrifuging. The amount of powder remaining in suspension after centrifugation expressed as a percent of the total amount in suspension is taken as the measure of solubility.

Apparatus

Boiling tube, centrifuge tube, aluminium dish, boiling water bath, Refrigerator, desiccator, weighing balance.

Procedure

1. Weigh accurately 4g of milk powder into a 50 ml boiling tube.
2. Add 32 ml of water at 50±1°C. Cork the tube and shake for 12 seconds.
3. Place the tube in a water bath at 50±1°C for 5 min.
4. Shake the tube for 1 min, making about 4-6 double excursions of about 30 cm per second.
5. Fill up to the brim, a 25 ml centrifuge tube with reconstituted milk and centrifuge for 10 min at 2000 rpm.
6. Cool in a refrigerator or in ice until the fat becomes solid and forms a cake, taking care that the milk does not freeze.
7. Remove the fat layer with as little as milk as possible, by running a needle around the cake of fat and then remove it with a spoon-shaped spatula. Warm the milk to 27±1°C.
8. Break up the deposit with a rod or wire. Cork the tube and sake well until the liquid appears to be homogenous.
9. Transfer about 2 ml of the homogenous liquid to a previously dried and weighed aluminium dish.
10. Weigh the dish with the liquid.
11. Re-centrifuge the tube for 10 min and pipette off about 2 ml of the upper layer of the supernatant liquid without disturbing the sediment, into a second aluminium dish and weigh.
12. Place both the dishes in a hot air oven at 100±1°C for 50 min. Transfer the dishes to a desiccator. Cool and weigh the dishes individually.

Observation

W_1 = weight in g of total solids in supernatant liquid

W_2 = weight in g of the liquid taken immediately after the removal of fat

W_3 = weight in g of total solids of the liquid taken immediately after the removal of fat

W_4 = weight in g of the supernatant liquid

Calculation

$$\text{Solubility (\%, by weight)} = \frac{W_1 \times W_2}{W_3 \times W_4} \times 100$$

6 Determination of Total Ash

Introduction

Ash content in milk powder is represents the inorganic residues remained after the organic matter has been burnt away at 550 ±10ºC temperature using a muffle furnace.

Apparatus

Platinum or silica crucible, muffle furnace, desiccator, weighing balance

Procedure

1. Heat the crucible in order to remove any moisture from it, cool in a desiccator and weigh.
2. Weigh accurately about 3 g of milk powder in above crucible. Place the crucible over a Bunsen burner and heat gently until the milk powder turn black.
3. Remove the crucible from the Bunsen burner and keep the crucible along with sample in a muffle furnace at a temperature not more than 550ºC until the ash is free from carbon.
4. Cool in a desiccator and weigh quickly.

Observation

W = weight in g of empty crucible

W_1 = weight in g of the crucible with milk powder

W_2 = Weight in g of the crucible after ashing

Calculation

$$\text{Ash, percent by weight} = \frac{(W_2 - W_1)}{(W_1 - W)} \times 100$$

7 Determination of Fat in Milk Powder

7.1 Gerber Method

Introduction

The sulphuric acid dissolves the protein and other milk solids and creates heat which liquefies and de-emulsifies the fat. Iso-amyl alcohol is added to aid the separation of fat.

Apparatus

Milk butyrometer, centrifuge machine, water bath at 65°C, weighing balance

Reagents

- Gerber sulphuric acid - 90% (Sp. gravity - 1.007-1.812 at 20°C)
- Iso-amyl alcohol – Specific gravity 0.803 to 0 .805 at 27°C

Procedure

1. 2.5g of milk powder is weighed accurately and added carefully to the 10 ml of concentrated sulphuric acid (sp. gr. 1.82) in a milk powder butyrometer.
2. Add little water over the sulphuric acid layer to avoid charring of milk powder.
3. All the steps are same as for the method (A.1.3.1) of fat estimation in milk by Gerber method.

7.2 Mojonnier Method

Introduction

The milk sample is treated with ammonia and ethyl alcohol; the former to dissolve the protein and the latter to help precipitate the proteins. Fat is extracted with diethyl ether and petroleum ether. Mixed ethers are evaporated and the residue weighed. This method is considered suitable for reference purposes. Strict adherence to details is essential in order to obtain reliable results.

Apparatus

Mojonnier tube, conical flask, hot water bath, weighing balance

Reagents

- Ammonia - Sp. gravity. 0.8974 at 16°C
- Ethyl alcohol - 95%

- Diethyl ether – Sp. gravity – 0.720, peroxide free
- Petroleum ether – Sp. gravity-0.65, Boiling point - 40-60°C

Procedure

1. Weigh accurately 1g of whole milk powder or 1.5g of partly skimmed milk powder into an extraction tube.
2. Add 10 ml water and shake until the milk powder is completely dissolved. Add 1.5 ml of ammonia solution (25%, w/v) and heat in a water bath (60-70ºC) for 15 min, shaking occasionally.
3. Cool and add 10 ml of ethyl alcohol.
4. Add 25 ml of each petroleum ether and di ethyl ether and mix and keep undisturbed for 30 min
5. Decant the clear mix ether layer into a conical flask
6. The above step is followed another two times with 15 ml of each petroleum and diethyl ether.
7. Collect all three-ether layer into the conical flask
8. Evaporate the ether using a hot water bath (boiling) and keep in a hot air oven (100ºC) for 1 hr.
9. Cool in a desiccator and take the weight

Observation

W_1 = Weight of sample taken

W_2 = Weight of empty dish

W_3 = Weight of dish with dry fat residue

Calculation

$$\%Fat = \frac{W_3 - W_2}{W_1} \times 100$$

8 Estimation of Scorched Particles (ADMI)

Introduction

These occur as unsightly, discolored specks in milk powders. They are often the result of powder deposits in the spray drying system. The amount of scorched particles in a powder is determined by comparison with the ADMI chart: 'Scorched Particle Standards for Dry Milk'. This method is used for milk powder and all other dried dairy products.

Apparatus

Blender, Standard cotton disc, aspirator

Procedure

1. Add 250 ml of sediment free water in blender and add 25g of skimmed milk powder or buttermilk powder, or 32.5g of whole milk powder.
2. Add approximately 0.5 ml of defoaming agent (diglycol laurate) and mix for 60 seconds in the blender.
3. Filter the entire solution through a standard cotton disc, using an aspirator.
4. Rinse the mixing container and tester with approximately 50 ml of sediment free water, also passing this through the cotton disc. Stir the re-liquefied sample thoroughly just before pouring it into the tester.
5. Remove the filter disc, place it in a scorched particle disc test card and dry at 30-40°C in a dust free atmosphere.
6. Compare the dry disc with the ADMI scorched particles standard photo print under uniform indirect light. Any test falling between two standard disc should be assigned the higher disc's letter.

6

Detection of Adulteration in Milk

1 Detection of Starch

Introduction

Starch is widely available in several forms such as wheat flour, corn flour and commercially manufactured starch. It is extensively used by adulterators to raise the SNF content in the milk. When high amounts of starch are added to milk, it can cause diarrhea due to the indigestion of starch in colon. Apart from this, accumulation of starch in the body may prove very fatal for diabetic patients. In this method, starch in milk can be detected by using iodine solution. Iodine from the iodine solution gets absorbed on the starch molecule surface and resultant product of adsorption has deep blue colour.

Reagents

- Potassium iodide
- Iodine solution - 1% (Add 1 g of iodine and 5 g of potassium iodide in distilled water and dilute to 100 ml)

Procedure

1. Take about 3 ml of well mixed milk in test tube.
2. Boil the milk in water bath.
3. Cool the boiled milk.
4. Add 2-3 drops of 1% iodine solution and shake well.

Observation

Appearance of blue color indicates the presence of starch in the milk sample whereas a control milk sample develops slight yellowish color.

2 Detection of Cane Sugar

Introduction

Cane sugar is widely used as adulterant by unscrupulous milk producers to increase the solids-not-fat (SNF) content of milk i.e. to increase the lactometer reading of milk, which was previously diluted with water. Cane sugar is quite cheap as compared to costly milk solids; therefore, adulterating milk with cane sugar will get the unscrupulous milk producer more profit. As per the FSSAI standards guideline of milk, it is a punishable offence when anything is added to the milk and, therefore detecting cane sugars presence in the milk samples is very essential. This test is specific for aldose sugar and not applicable for ketose sugar. Hence, sucrose being ketose can be detected by using this test. Sucrose contains fructose a keto sugar, which is released upon hydrolysis by reaction with HCl. Ketoses undergo dehydration to give furfural derivatives which condenses with resorcinol to form red coloured condensation products.

Reagent

- Resorcinol powder
- Concentrated Hydrochloric Acid (HCl)

Procedure

1. Take about 5 ml of milk in a dry clean test tube.
2. Add 1 ml of Concentrated HCl.
3. Add 0.1 g of resorcinol and mix content well.
4. Place the tube in boiling water bath for 5 min.

Observation

In the presence of cane sugar in the milk sample, red color is produced. Whereas in pure milk samples no such red color is develop and sample remains white with visible coagulum in it.

3. Detection of Skim Milk Powder

Introduction

As per the law, skimmed milk powder (SMP) cannot be used for adjustment of SNF content of cow/buffalo or mixed milk. Though, SMP can be used for the modification of milk solids in case of toned, double toned and recombined milk. This method is based on the fact that the coagulum obtained from reconstituted skim milk powder by addition of acetic acid, gives intense blue color. Intense blue colour is further formed on boiling with phosphomolybdic acid due to certain reducing groups present in the proteins of milk powder which are able to cause reduction of molybdenum.

Reagent

- Diluted acetic acid: 4% (v/v).
- Phosphomolybdic acid solution: 1% (w/v) solution in distilled water

Procedure

1. Take 50 ml of milk in a 60 ml centrifuge tube.
2. Place the tube in the centrifuge and centrifuge at 3000 rpm for 15 minutes.
3. Pour the supernatant creamy layer carefully.
4. Add 0.5 ml of diluted acetic acid for coagulation and then add 2 ml of phosphomolybdic acid solution.
5. Mix the contents thoroughly and heat it in a water bath at boiling temperature for 15 minutes and then cool at room temperature.
6. Observe the colour of the curd obtained.

Observation

The curd of the sample containing skimmed milk powder shall be bluish in colour whereas, the curd obtained from pure milk shall be greenish in colour. The blue colour intensity depends on the amount of the skim milk powder present in the sample.

4 Detection of Urea in Milk

Introduction

Urea is a native compound of milk and it represents a major part of the non-protein nitrogen of milk. Urea is widely used in synthetic milk along with caustic soda, detergent, sugar and foreign fats. Adulteration of synthetic milk with natural milk leads to increase in concentration of urea which causes toxicological hazards. Estimation of urea concentration in milk may serve as a tool for checking the menace of adulteration of natural milk with synthetic milk. In cow milk, average urea content was found in the range of 20 to 35 mg/100 ml. whereas, in buffalo milk, it was found in range of 25 to 40 mg/100 ml (Kavita, 2000). As per FSSAI Act 2006 and PFA Rules (1955), the urea content in milk should not be more than 70 mg/ 100 ml. The addition of urea to milk can be detected by using p-dimethylaminobenzaldehyde(DMAB). This method is based on the principle that urea forms a yellow complex with DMAB in a low acidic solution at room temperature. The intensity of yellow color depends on the concentration of urea present in the milk.

Reagent

- Dimethyl amino benzaldehyde (DMAB): Mix 1.6 g of DMAB in Ethyl alcohol containing 10 ml of concentrated HCl.
- Ethyl alcohol
- Concentrated HCI

Procedure

1. Take 1 ml of milk in a test tube.
2. Add 1 ml of 1.6% (w/v) DMAB reagent and mix content well.
3. Note the color obtained.

Observation

The formation of distinct yellow color indicates the presence of added urea in milk sample whereas slight yellowish color develops in control milk sample as pure milk itself contains 25-40 mg/100ml urea.

5 Detection of Ammonium Sulphate

Introduction

Ammonium Sulphate is added to the milk as it increases the lactometer reading by maintaining the density of milk. Addition of sulphates in milk increases the lactometer reading and it also raises the SNF content fraudulently. Several ammonium salts e.g. ammonium chloride, ammonium sulfate, ammonium nitrate and ammonium dihydrogen orthophosphate can be detected by adding sodium hydroxide, sodium hypochlorite and phenol. The reaction of the ammonium sulphate with three reagents leads to formation of deep blue colour. The deep blue color is generated when the amine reacts with phenol in the presence of hypochlorite in an alkaline environment, results in the formation of a complex which is blue in color.

Reagents

- Sodium hydroxide solution: 2% (w/v) in distilled water
- Sodium hypochlorite solution: 2% in distilled water
- Phenol solution: 5% (w/v) in distilled water

Procedure

1. Take 1 ml of milk in a test tube.
2. Add 0.5 ml of sodium hydroxide and 0.5 ml of sodium hypochlorite solution.
3. Mix thoroughly and to this adds 0.5 ml phenol solution.
4. Heat the tube in boiling water bath for 20 seconds.

Observation

Development of blue colour indicates the presence of extraneous ammonium sulphate in the milk.

6 Detection of Glucose

Introduction

Milk contains trace amount of glucose. Glucose is added in milk to raise the total solid content of milk. Glucose being a reducing sugar poses many problems in its detection because of lactose, major carbohydrate in milk, is also a reducing sugar. Moreover, it is easily available in commercial form as concentrated syrup. Added glucose in milk can be detected by using Barfoed's reagent. Barfoed's reagent uses copper in the acidic medium. Due to weakly acidic nature of Barfoed's reagent, copper is reduced only by monosaccharide. The test is selectively only for the monosaccharide as glucose which reacts faster (2-3 min) than disaccharide (7-10 min). Thus, glucose can be detected.

Reagents

- Barfoed's reagent: Add 24 g of cupric acetate in 450 ml distilled water (boiled water). Add 25 ml of 85% lactic acid to the solution immediately. Mix well and filter. Make the valume of the filtrate to 500 ml with distilled water.
- Phosphomolybdic acid: Mix 35 g of ammonium molybdate and 5 g of sodium tungstate in 200 ml of 10% (w/v) sodium hydroxide solution and 200 ml distilled water. Boil for 1 hr. Add 350 ml of water and add 125 ml of phosphoric acid (85%). Dilute to 500 ml.

Procedure

1. Take 1 ml of milk sample in test tube.
2. Add 1 ml of Barfoed's reagent.
3. Heat the tube in boiling water bath for 3 min and cool under tap water for 3 min.
4. Add 1 ml of phosphomolybdic acid reagent to the turbid solution.

Observation

Immediate formation of deep blue color after adding phosphomolybdic acid reagent indicates the presence of added glucose in the milk sample. In case of pure milk only faint bluish color can be observed.

7 Detection of Added Natural Water

Introduction

Pond water also contains appreciable quantities of nitrates such as sodium and potassium nitrates. Pond water is usually admixed with milk by rural milk producers or vendors for adjustment of lactometer reading by adding different adulterant. In the pond water nitrates may come from fertilizers used in the fields. This method actually detects nitrates present in the pond water. It can be easily detected by using diphenylamine and concentrated sulfuric acid reagent which produce blue colour in the presence nitrates.

Reagents

- Diphenylamine: 2% (w/v) in distilled water
- Concentrated Sulfuric acid

Procedure

1. Take 5 ml milk in a test tube, add 6-7 drops of 10% acetic acid and shake for 2 min.
2. Take 2 ml of diphenylamine reagent in another test tube.
3. Filter the content of the sample tube and collect about 1 ml filtrate over diphenylamine reagent.
4. Development of blue colour indicates presence of nitrate in milk and thus of added natural water.

Observation

Deep blue color will be formed in presence of nitrate in the milk sample. Pure milk sample will not develop any color.

Note: Glassware used in this test must be cleaned properly with nitrate free distilled water and dried.

8 Detection of Dextrin / Maltodextrin

Introduction

Maltodextrins are formed as intermediate products during hydrolysis of starch by dilute solution of acid or amylases. As compare to native starches, the maltodextrins are soluble in water and do not give characteristic blue color with iodine solution (Chronakis, 1998). In recent time, maltodextrin considered potent adulterant. It has been reported mainly to increase the lactometer reading or SNF content and also to increase the yield of the product prepared from it such as khoa and burfi. In this method, maltodextrin in milk can be detected by using dilute iodine solution (0.05 N). Iodine forms a coordinate complex between the helically coiled polysaccharide chains and get centrally located within the helix due to adsorption and develops chocolate brown colour. The colour obtained depends upon the degree of branching length of the polysaccharide chain available for complex formation.

Reagents

- Iodine solution (0.05 N)

Procedure

1. Take about 5 ml milk sample in a test tube.
2. Add 2 ml of dilute iodine solution (0.05N) to the tube and mix well.
3. Observe for the change development.

Observation

Development of chocolate-red brown color indicates the presence of maltodextrin in the milk sample whereas in pure milk sample slight yellowish color can be observed.

9 Detection of Neutralizers in Milk

Introduction

The natural acidity of freshly drawn milk found in the range between 0.12-0.16% (lactic acid) for cow as well as buffalo milk. It is mainly due to phosphate, casein, and to a less extent citrate and carbon dioxide. Upon prolonged storage, the milk bacterium grows and produces lactic acid which leads to spoilage of milk. So, to mask the developed acidity and to increase shelf life of milk, middlemen attempt to add neutralizers in the form of sodium carbonate, sodium bicarbonate, lime and sodium hydroxide. But such a practice of adding neutralizers to milk is not permissible under PFA Rules, 1955.

Rosalic acid test and alkalinity of ash methods can be used for the detection of neutralizers in milk. Rosalic acid is a pH indicator which shows a change in the color on addition to alkaline milk at pH 7.0 to 8.0. Addition of neutralizers in milk moves its pH towards alkaline side. Rosalic acid has red colour in the alkaline side. This test will work only when neutralizers are added in excess quantities and milk is alkaline in nature. If the sour milk is under-neutralized below the normal pH of the milk, then this test will fail to detect added neutralizers. The other test for the detection of neutralizers in milk i.e. test for alkalinity of ash is based on the fact that the neutralization of milk invariably increases the ash content and also the total alkalinity of the ash from a fixed quantity of milk.

9.1 Rosalic Acid Test

Reagents

- Rosalic acid (0.1%, w/v)
- Ethyl alcohol (95%)

Procedure

1. Take about well mixed 5 ml milk in a test tube.
2. Add 5 ml ethanol and mix the content well.
3. Add of 2-3 drops of Rosalie acid solution.
4. Note the occurrence of color in the test tube.

Observation

If carbonate or bicarbonate is present red colour will appear. If sodium hydroxide is present deep rose red colour will appear. Whereas pure milk samples produce brownish or brownish yellow color only.

9.2 Alkalinity of Ash

Addition of alkaline substances e.g. sodium carbonate, bicarbonate and sodium hydroxide will increase the alkalinity of ash and hence will give fair idea about neutralization of milk.

Reagents

- Concentrated Hydrochloric acid (HCl)
- Phenolphthalein indicator

Procedure

1. Take 20 ml of milk sample into a silica dish and dry it over boiling water bath.
2. Keep dried milk sample in Muffle Furnace at 550±50°C for 4-6 hour for the preparation of ash.
3. Let the ash residue cool to room temperature.
4. Disperse the ash in 10 ml water and titrate the ash content by standard 0.1N HCI using phenolphthalein indicator.

Observation

If the volume of 0.1N HCI required to neutralize the ash content exceeds 1.20 ml, then the milk is suspected to contain neutralizers.

10 Detection of Preservatives in Milk

Introduction

Several preservatives were added in milk to increase its shelf life such as hydrogen peroxide, formalin/formaldehyde, benzoic and salicylic acid etc. Although addition of any kind of preservative is not legally permitted in India, traders used to add various kinds of preservatives to milk.

10.1 Detection of Formaldehyde

Formalin (40 % aqueous solution of formaldehyde) is widely used for the preservation of milk and milk products to check their adulteration. Apart from laboratory purpose, addition of formalin to milk is legally prohibited as it is very poisonous chemical. Yet, market milk samples are occasionally found adulterated with formaldehyde to prevent the rise in acidity. Formalin or formaldehyde added milk forms characteristic violet colour with ferric salts and other oxidizing agents. Hehner and Lech tests are commonly used for the detection of formaldehyde/formalin in milk

10.1.1 Hehner Test

Procedure

- Take 10 ml of milk sample in a test tube.
- Add 0.5 ml of ferric chloride solution and mix well.
- Add concentrated sulphuric acid carefully and slowly from the side so that it forms separate layer at the bottom of the tube without mixing with milk.

Observation

Development of violet ring at the junction of two layers indicates presence of formaldehyde.

10.1.2 Leach Test

Procedure

- Take 5 ml of milk in a test tube
- Add carefully equal volume of hydrochloric acid containing ferric chloride.
- Heat the tube for 5 min in light temperature.
- Rotate the tube to break the curd and observe for the colour.

Observation

Appearance of violet colour in the curd indicates presence of formaldehyde.

10.2 Detection of Hydrogen Peroxide

Procedure

- Take 5 ml of milk in test tube
- Add 2 drops of p-phenyldiamine hydrochloride solution and mix well.

Observation

Appearance of intense blue colour in the tube indicates presence of hydrogen peroxide.

10.3 Detection of Carbonate and Bicarbonate

Procedure

- Take 5 ml milk sample and 5 ml of ethyl alcohol in the test tube.
- Add few drops of rosalic acid solution and mixed well.

Observation

Development of rose red colour indicates presence of carbonate or bicarbonate.

10.4 Detection of Benzoic and Salicylic Acid

Procedure

- Take 5 ml of milk in a test tube.
- Add concentrated sulphuric acid carefully for acidification of milk.
- Add drop by drop 0.5% ferric chloride solution and mix the content well.

Observation

Development of buff colour and violet colour indicates presence of benzoic acid and salicylic acid respectively.

10.5 Detection of Borax and Boric Acid

Procedure

- Take 5 ml of well mixed milk sample in a test tube.
- Add 1 ml concentrated hydrochloric acid and mix the content well.
- Dip the tip of turmeric paper in the acidified milk.
- Allow to dry the paper in watch glass at 100°C or over a small flame.

- After drying, if turmeric paper turns red it shows the presence of borax or boric acid.
- Add a drop of ammonia solution on the same turmeric paper.

Observation

Change of red colour of turmeric paper to green indicates presence of boric acid.

7

Preparation of Standard Solutions

1. Preparation of Standard Sodium Hydroxide Solution

Introduction

When oxalic acid reacts sodium hydroxide, sodium oxalate and water are form. Phenolphthalein indicator detects the end point by its characteristics colour change.

Apparatus

Burettes, pipette, volumetric flask, beaker

Reagents

- Oxalic acid crystals
- Sodium hydroxide – 0.1 N approximate
- Phenolphthalein solution – 0.1 percent in ethyl alcohol

Procedure

Preparation of 0.1 N oxalic acid solution

1. Weight accurately 1.575 g of oxalic acid into a 250 ml beaker using a weighing bottle.
2. Add some distilled water to the beaker and dissolve the oxalic acid with the help of glass rod.
3. Transfer the solution into a 250 ml volumetric flask through a funnel.
4. Wash the beaker 3- 4 times with small quantities of distilled water and transfer the washing into the flask.
5. Make up the volume up to the mark with distilled water and mix thoroughly.
6. Preparation of approximately 0.1 N sodium hydroxide solution
7. Weigh about 4.25 g of sodium hydroxide pellets and dissolve into 200 ml of distilled water.

8. Shake the mixture and invert three to four times.
9. Standardization of sodium hydroxide solution:
10. Take 25 ml of aliquot portion of oxalic acid in a conical flask.
11. Add 3 – 4 drops of phenolphthalein solution and mix the content.
12. Fill the burette with sodium hydroxide,
13. Add sodium hydroxide drop wise from the burette into the conical flask till a permanent pale pink colour appears and persists for 30 sec in the solution.
14. Repeat the experiment three times more and take the average reading.

Calculation

Normality of oxalic acid solution $= \dfrac{(a-b)\times 0.1}{1.575}$

Where,

a = Weight in g of weighing bottle before transferring oxalic acid

b = Weight in g of weighing bottle after transferring

Normality of sodium hydroxide solution (N2) $= \dfrac{V1N1}{V2}$

Where,

V1 = Volume in ml of oxalic acid solution taken

V2 = Volume in ml of sodium hydroxide solution required for titration

N1 = Normality of oxalic acid solution

N2 = normality of sodium hydroxide solution

2. Preparation of Standard Hydrochloric Acid Solution

Introduction

Solution of sodium carbonate is titrated against hydrochloric acid in presence of methyl orange as an indicator.

Apparatus

Weighing balance, burette, pipette, conical flask, volumetric flask, funnel, weighing boat

Reagents

- Anhydrous sodium carbonate,

- Hydrochloric acid,
- Methyl orange indicator

Procedure

Preparation of 0.1 N sodium carbonate solution

1. Take 1.325 g of sodium carbonate in a 250 ml beaker.
2. Add some distilled water into the beaker and dissolve the sodium carbonate with the help of a glass rod.
3. Transfer the solution in a 250 ml volumetric flask.
4. Wash the beaker 3 - 4 times with small amount of distilled water and transfer the washing into volumetric flask.
5. Make up the volume with distilled water up to the mark and mix thoroughly.

Preparation of an approximately 0.1 N hydrochloric acid solution

1. Measure 9 ml of concentrated hydrochloric acid in 1 lit volumetric flask containing 400 – 500 ml distilled water
2. Make the volume up to the mark with distilled water.
3. Insert the stopper and thoroughly mix the solution by shaking and inverting the flask repeatedly.

Standardization of hydrochloric acid solution

1. Take a clean dry burette and rinse it 2 times with hydrochloric acid solution.
2. Fill it completely up to the zero mark with hydrochloric acid solution and fix it in a burette stand.
3. Pipette out 25 ml of the sodium carbonate solution into a 250 ml conical flask.
4. Add one to two drop of methyl orange and mix the content of the solution.
5. Add the acid solution from the burette drop wise into alkali solution until it appears a permanent orange colour.
6. Repeat the experiment three times more.
7. Take the average of the concordant results (difference not over 0.1 ml).

Calculation

$$\text{Normality of sodium carbonate solution} = \frac{(a-b)\times 0.1}{1.325}$$

Where,

a = Weight in g of weighing bottle with sodium carbonate before transferring

b = Weight in g of weighing bottle with sodium carbonate after transferring

Normality of hydrochloric acid solution: (N2) $= \frac{N1 \times V1}{V2}$

Where,

V1 = Volume in ml of sodium carbonate solution taken

V2 = Volume in ml of hydrochloric acid solution taken for titration

N1 = Normality of sodium carbonate

N2 = Normality of hydrochloric acid

3. Preparation of Standard Silver Nitrate Solution

Introduction

The method is based on titration of sodium chloride with silver nitrate solution in presence of potassium chromate as indicator. The end point of the titration is the point at which the colour of the suspension changes from pure yellow to reddish brown. Due to presence of CrO_4- ion in solution precipitation of red insoluble silver chromate occurs.

Apparatus

Burette, pipette, volumetric flask, funnel, beaker

Reagents

- Sodium chloride
- Silver nitrate: 0.1 N (approx)
- Potassium chromate solution – 5 % in distilled water

Procedure

Preparation of 0.1 N sodium chloride solutions

1. Weigh accurately about 1.461 g of pure sodium chloride into a beaker.
2. Add small quantity of distilled water and dissolve the sodium chloride in the beaker with a glass rod.
3. Transfer the solution into a 250 ml volumetric flask using a funnel.
4. Rinse the beaker 2 – 3 times with small quantity of distilled water and pour into the flask.

5. Make up the volume of the flask p to the mark with distilled water, Stopper it and thoroughly mix by inverting 3 – 4 time

Preparation of 0.1 N solution of silver nitrate

1. Take a clean burette and rinse it with the silver nitrate solution.
2. Fill the burette with silver nitrate solution and fix it in a burette stand.
3. Pipette out 25 ml sodium chloride solution in 250 ml conical flask.
4. Add 0.5 – 1 ml potassium chromate solution and shake the flask.
5. Add silver nitrate solution from the burette drop wise till the precipitate just begins to appear, reddish brown on colouration.
6. Repeat the precise titration once or twice more and take the average reading.

Calculation

Normality of silver nitrate solution $(N1) = \frac{N2 \times V2}{V1}$

Where,

V1 = Volume in ml of silver nitrate solution required by 25 ml of sodium chloride solution

N1 = Normality of silver nitrate solution

V2 = Volume in ml of sodium chloride taken

N2 = Normality of sodium chloride solution

4. Preparation of Standard Sodium Thiosulphate Solution

Introduction

Standardization of sodium thiosulphate is based on the general principle of iodometric determination of oxidizing agents. First a measured volume of standard potassium dichromate solution is added to a mixture of potassium iodide and hydrochloric acid. The dichromate is then replaced by an equivalent amount of free elemental iodine which is titrated with the thio sulphate solution to be standardised.

$$K_2Cr_2O_7 + KI + HCl \longrightarrow KCl + CrCl_3 + H_2O + I_2$$

Apparatus

Burette, pipette, volumetric flask, beaker

Reagents

- Sodium thio sulphate crystal
- Potassium dichromate crystal
- Concentrated hydrochloric acid
- Sodium carbonate

Procedure

Preparation of 0.1 N sodium thiosulphate solution

1. Weigh accurately 6.2 g of sodium thiosulphate crystal in a beaker.
2. Add some distilled water into the beaker and dissolve the sodium thiosulphate.
3. Transfer the solution into a 250 ml volumetric flask.
4. Add small quantities of distilled water into the beaker and transfer the washing into the volumetric flask.
5. Make up the volume up to the mark with distilled water.

Preparation of 0.1 N potassium dichromate solution

1. Weigh accurately 1.2257g potassium dichromate on an analytical balance and transfer it quantitatively into a 250 ml volumetric flask.
2. Make up the volume to the mark with distilled water and mix thoroughly.

Standardization of sodium thiosulphate solution

1. Take 100 ml of distilled water in 500 ml conical flask.
2. Add 15 ml of 20% potassium iodide solution and 2 gm of sodium carbonate.
3. Mix thoroughly.
4. Add 6 ml of concentrated hydrochloric acid with gentle rotation of the solution.
5. Add 25 ml standard solution of potassium dichromate solution from a pipette, cover the flask with a watch glass to prevent losses due to volatilization of iodine and leave the mixture for 5 min in a dark place.
6. Now remove the watch glass and rinse it with distilled water in the flask.
7. Add about 200 ml of distilled water.
8. Fill the burette with sodium thiosulphate solution and adjust the liquid level to the zero mark.

9. Titrate the liberated iodine with the thiosulphate solution taken in the burette with constant circular movement of the mixture in the conical flask.
10. At first titrate without indicator.
11. When the colour of the solution has changed from dark brown to pale yellowish (straw yellow), add 2 ml of starch solution
12. Continue the titration until a single drop of sodium thiosulphate solution changes the colour blue to light green.
13. The end point becomes sharp against a white background.
14. Repeat the experiment two times more.
15. Take the average of the concordant results (difference not over 0.1 ml).

Calculation

Normality of potassium dichromate solution $= \frac{(a-b)\times 0.1}{1.2257}$

Where,

a = Weight in g of potassium dichromate before transferring

b = Weight in g of potassium dichromate after transferring

Normality of sodium thiosulphate solution (N1) $= \frac{V2\times N2}{V1}$

Where,

V1 = Volume in ml of sodium thiosulphate solution required for titration

V2 = Volume in ml of potassium dichromate solution taken for the test

N1 = Normality of sodium thiosulphate solution

N2 = Normality of potassium dichromate solution

8

Calibration of Glassware

1. Calibration of Milk Butyrometer

Introduction

Butyrometer is used for determination of fat for milk and milk products by Gerber method. The milk butyrometer is a graduated vessel, which are designed in such a way that the internal volume of each unit is 0.125 ml which is corresponding to 1 percent fat. Sometimes due to manufacturer's error, the volume of one unit of milk butyrometer is not exactly equal to 0.125 which leads to subsequent error in results in milk fat. So before starting to use a new butyrometer, it is essential to calibrate for its efficiency in the laboratory. Mercury is used to calibrate milk butyrometer.

Apparatus

Milk butyrometer (0 – 10% graduation), analytical balance

Reagents

- Mercury

Procedure

1. Clean the butyrometer with dilute sodium hydroxide, tap water, chromic acid and distilled water to make it free from grease and dust.
2. Dry the butyrometer by rinsing with acetone then under cold dry air and finally tag with a wire loop at the neck to facilitate weighing.
3. Fill the butyrometer with mercury using a micropipette up to the graduation mark nearest to the closed end (10 percent mark),
4. Adjust the meniscus with the help of magnifying glass.
5. Weigh the milk butyrometer
6. Fill again to the next graduation mark (9 percent) and weigh
7. Repeat filling and weighing procedure till the mercury reaches to zero percent graduation.

8. From the weight and density of mercury at room temperature, internal volume of milk butyrometer can be calculated.
9. The milk butyrometer which conform to the calculated volume of 0.125 ± 0.002 ml may be used to determination of fat otherwise rejected.

Calculation

In milk butyrometer,

$$\text{Volume of 1\% graduation} = \frac{(W_2 - W_1)}{D(10 - x)}$$

Where,

W_1 = Weight in gm of milk butyrometer with mercury up to 10% graduation mark

W_2 = Weight in gm of milk butyrometer with mercury up to x % graduation mark

D = Density of mercury (13.6 gm/ml at 20°C)

x = Graduation mark in % in milk butyrometer

2. Calibration of Milk Pipette

Introduction

Milk pipettes are commonly used for accurate measurements of definite volume of milk. They are long narrow tubes with a bulb in the middle. The upper narrow part of the pipette has a line, up to that the pipette is filled. The volume of milk pipette is determined by the volume of distilled water at 27°C, expressed in terms of ml delivered by the pipette. The volume so determined shall be within the following limits.

Milk pipette = 10.75 ± 0.03 ml

Sulphuric acid pipette = 10.25 ± 0.05 ml

Amyl alcohol = 1.00 ± 0.05 ml

Apparatus

Milk pipette (10.75 ml), analytical balance and beaker

Procedure

1. Clean the pipette with detergent, tap water, chromic acid, tap water and distilled water respectively and dry it.
2. Fill the pipette with distilled water about 2 cm above the mark.

3. Close the top of the pipette quickly with index finger.
4. Wipe out the distilled water from the outside portion of the lower end of the milk pipette with a piece of clean filter paper.
5. Place a tarred 50 ml capacity beaker below the pipette and discharge the distilled water from thc pipette slowly until the bottom edge of the meniscus just touches the mark.
6. Determine the weight of distilled water thus delivered.
7. Repeat the experiment at least three times.
8. Take the average reading and note down the temperature of the distilled water.

Calculation

Volume in ml of milk pipette = m/d

Where,

m = weight in g of distilled water

d = density of distilled water at room temperature

3. Calibration of Lactometer

Introduction

Lactometer, which works on the Archimedes principle, is basically a specific gravity (sp. gr.) hydrometer specifically designed for milk. Stem of the lactometer has graduation range. Lactometers are calibrated by floating it in the solution of pure anhydrous sodium chloride of desired concentration exhibits the definite specific gravities at the calibration temperature.

Apparatus

Lactometer, beaker, thermometer

Reagents

- Sodium chloride

Procedure

1. Prepare two solution of pure sodium chloride of different concentrations of 3.863 % and 4.415 % by dissolving it in distilled water.
2. Determine the specific gravity at specified temperature (eg. Quevenne lactometer at 15.5°C or 60°F by using specific gravity bottle or pycnometer).

3. Record the lactometer reading at 60°F with the lactometer to be tested by floating in these solutions with the same manner as is done in case of milk.
4. Compare the accuracy of the lactometer in all the solutions (Table 4).

Table 4: Lactometer reading in different solutions:

Sodium chloride solution (%)	**Specific gravity at 15.5°C (g/ml)**	**Lactometer reading at 15.5° C**
3.863	1.026	26
4.415	1.032	32